DISCOVERY OF GREEN FORMOSA

發現綠色台灣

台灣植物專輯

中華民國企業永續發展協會會員名單	**Member List of BCSD-Taiwan**
中美和石油化學股份有限公司	China American Petrochemical Co.
中鼎工程股份有限公司	CTCI Corporation
中國國際商業銀行	The International Commercial Bank of China
中華開發工業銀行股份有限公司	China Development Industrial Bank
中興紡織廠股份有限公司	Chung Shing Textile Co.
永豐餘造紙股份有限公司	Yuen Foong Yu Paper Mfg. Co.
台灣水泥股份有限公司	Taiwan Cement Corporation
台灣永光化學工業股份有限公司	Everlight Chemical Industrial Corp.
台灣苯乙烯股份有限公司	Taiwan Styrene Monomer Corporation
台灣積體電路製造股份有限公司	Taiwan Semiconductor Manufacturing Co.
東元電機股份有限公司	TECO Electric & Machinery Co.
英業達股份有限公司	INVENTEC Corporation
南聯國際貿易股份有限公司	Nanlien International Corporation
財團法人台灣電子檢驗中心	Electronics Testing Center Taiwan
財團法人祐生研究基金會	Archilife Research Foundation
財團法人環境與發展基金會	Environment and Development Foundation
統一企業股份有限公司	Uni-President Enterprise
統一超商股份有限公司	President Chain Store Corporation
寶成工業股份有限公司	Pou Chen International Group

發行單位 Publishers

行政院農業委員會林務局・社團法人中華民國企業永續發展協會

Taiwan Forestry Bureau, Council of Agriculture, Executive Yuan, R.O.C ・ Business Council for Sustainable Development, R.O.C

贊助單位 Sponsors

行政院國家永續發展委員會・行政院環境保護署

Executive Yuan's National Council for Sustainable Development ・ Environmental Protection Administration ・ Executive Yuan, R.O.C

中美和文教基金會・中國石油公司・台積電文教基金會

CAPCO Cultural & Educational Foundation ・ Chinese Petroleum Corp. ・ TSMC Education and Culture Foundation

2002年6月

發現綠色台灣
Discovery of Green Formosa

C 目 O 錄 N T E N T S

序一

本會在1998年和2000年結合各界的努力，相繼製作了「水藍之舞一綠島海洋資源保育散文攝影專輯」與「消失中的精靈一台灣珍貴及稀有動物保育專輯」兩本針對生物多樣性議題的專書，來紀錄台灣本土的生物多樣性，頗獲國內外各界的好評。因此在這動力的驅使下，深感有必要繼海洋生物、動物之後，製作我國珍貴的植物保育專輯，以擴大對國內外保育意識之倡導。

赤道地區是人口增長最快的區域，號稱「地球之肺」的赤道雨林，每年被砍伐了二到三個台灣大小的面積，因此功能逐漸衰退，導致全球氧、碳的循環逐漸遲緩，進而造成全球暖化、氣候異常等情況越來越嚴重。此外，工業的廢氣排入空氣中，和水滴結合成飽含酸度的酸雨，亦會摧殘整片森林。在失去森林的庇蔭後，許多種類的生物也將迅速從地球上消失。因此如何善用地球有限的資源，維護人類的生活環境，增進人類的福祉，將是每一位地球人的責任。

本書集合了國內植物界二十幾位精英的專業知識，及各界對植物保育的熱誠，以通俗的文字與生動的照片，將台灣美麗及寶貴的植物，介紹給社會大眾及國外人士，進而達到宣導及教化的成效。

本人在此謹代表本會向諸位不辭勞苦的作者們致上萬分謝意。其次是感謝台大植物標本館、植物系郭教授城孟及其助理們的大力幫忙，使得本書能在如此短的時程內完成。

最後，感謝行政院國家永續發展委員會、行政院農業委員會林務局、環保署、台積電文教基金會、中美和文教基金會、及中國石油公司的支持與贊助，使本書得以順利出版。這是繼「消失中的精靈一台灣珍貴及稀有動物保育專輯」後，政府部門及民間團體再次攜手合作，共同為台灣資源保護的工作盡份心力之另一典範。

社團法人中華民國企業永續發展協會　理事長

陳耀生 謹識

Foreword 1

With the combined efforts of people from various fields, from 1998 to 2000 the Council has continued production of two books on biodiversity topics: "Blue Dancers" - Photographs and Essays on the Conservation of Marine Resources by the Green Island and "Vanishing Dancers" - Conservation of Taiwan's Precious Rare Animal Species. These books are meant to record the native biodiversity of Taiwan. They have been very favorably received both in Taiwan and abroad. With this impetus, we feel deeply the necessity to continue the work beyond the marine and animal areas to produce a conservation work addressing Taiwan's precious plant resources. This will serve to broaden domestic and foreign consciousness about conservation.

The tropics are the region of fastest growth in human population, and the rain forest, which have been called "the lungs of the earth" are being cut down at the rate of an area two to three times the land area of Taiwan each year. Their function is weakening as a result. This has led to a gradual slowdown in the global oxygen and carbon cycles and created daily worsening global warming, climate abnormalities and other phenomena. Additionally, industrial wastewater and emissions and acid rain have also accelerated the damage to our forests. As the forests disappear, many biodiversity species are likewise fast disappearing from the face of the globe. Consequently, how to conserve the earth's limited resources and save the living environment of humankind while increasing human welfare will be the responsibility of every global citizen.

This book collects outstanding professional knowledge from about twenty expertises in Taiwan, and efforts by all sectors of society about conservation of plant species, and using common language and lively photographs to introduce Taiwan's beautiful and precious plants to the society and to people abroad, achieving the goal of dissemination and education.

On behalf of the Council, I would like to express deep gratitude toward all the authors for their unstinting efforts, and also to thank the National Taiwan University Herbarium, Professor Chen-Meng Kuo and his assistants for the hard work that made the completion of this book in such a short time possible.

Finally, I would like to thank the Executive Yuan's National Council for Sustainable Development,Environmental Protection Administration, the Taiwan Forestry Bureau of the Council of Agriculture of the Executive Yuan, the TSMC Education and Culture Foundation, the CAPCO Cultural and Educational Foundation, and Chinese Petroleum Corp. for their support and assistance. in helping bring this book to publication. After " Vanishing Dancers", this book demonstrates that the government and private sector have once again joined hands to do their best on yet another effort for the conservation of Taiwan's nature resources.

Yao-Sheng Chen
Chairman
BCSD-Taiwan, R.O.C

序二

台灣是受板塊運動而於亞洲大陸棚東緣低緯帶所形成之大陸性島嶼，過去曾長期與大陸相連，於冰河時期成為亞洲大陸生物遷徙的避難所。因受多次造山運動影響，地形錯綜複雜，高低差近四千公尺，又有北迴歸線通過以及四周海洋的調節，因而造就了台灣豐富的動植物相。隨著海拔高度呈現熱帶、溫帶、寒帶三型氣候垂直分布，同時出現熱、暖、溫、寒帶性的不同動物及植物種類。台灣全島面積僅三萬七千七百八十平方公里，卻孕育著豐富多樣化的自然資源，我們除了欣賞不破壞外，更應積極負責地將它保留給後代子孫。

為了喚醒國內外人士對於本土珍貴動植物資源的重視與珍惜，當企業永續發展協會倡議編製「發現綠色台灣」這本書時，環保署即全力支持。本書由行政院國家永續發展委員會、農委會林務局與中華民國企業永續發展委員會合作，邀請國內植物專家及學者，共同紀錄台灣珍貴的生物資源，然後彙集成冊，可說是政府、企業與學術研究精英聯手奉獻的典範。透過專家學者深入淺出的筆觸，讓讀者更認識台灣珍貴、特有及稀有的植物，也更萌生珍惜之心。今年2002年8月「世界永續發展高峰會」即將在南非約翰尼斯堡舉行，我們希望藉由這本書的出版，能重現福爾摩莎的綠色之美，並喚起各界群策群力加入保護台灣自然資產的行列。

行政院環境保護署　署長

郝龍斌 謹識

Foreword 2

Human civilization is confronting a tough challenge in the 21st Century. In the face of environmental crises and degradation, we now must rethink about the relationship between man and nature and the delicate balance between development, environmental protection, and ecological conservation. The challenges these days are no longer about reconciling differences of opinions, but making the right decision in life-and-death situations.

Taiwan's ecology is complex and variegated, which makes it all the more precious. As Taiwan transformed from a backward agricultural society to an industrialized country, all the negative impacts arisen from the struggle between environmental protection and economic development have befallen this land. This is everyone's problem, and everyone is responsible. As we lament the deteriorating quality of life, let's also ponder on how we can contribute to protect our environment.

As a way to underscore the importance and value of Taiwan's precious plant resources, the Business Council for Sustainable Development has put together Discovery of Green Formosa. The Environmental Protection Administration is pleased to give its full support to this effort. This volume is the result of collaboration between the Nation Council for Sustainable Development, the Taiwan Forestry Bureau of the Council of Agriculture of the Executive Yuan and the Business Council for Sustainable Development. This volume has great significance in that it not only is a prime example of collaborative work between the government, the business sector, and academia, but its lucid writing offers easy-to-understand explanations of very weighty and complex topics, giving the reader a better awareness of the value, uniqueness and preciousness of Taiwan's plants. The World Summit on Sustainable Development Summit will be held in Johannesburg, South Africa in August 2002, and we hope the publication of this book will present the green beauty of Formosa and inspire people in all fields to join together for the protection of Taiwan's natural resources.

Lung-Bin Hau
Administrator
Environmental Protection Administration, Executive Yuan, R.O.C

序三

生物多樣性保育在世界各國所受到的重視程度，可由生物多樣性公約迄今已有180個國家或經濟體簽約成為締約國，可見一斑。在國內，生物多樣性保育推動層級已提昇至行政院，民國九十年八月行政院第二七四七次會議通過「生物多樣性推動方案」，自九十年七月起由各部會分工執行，將生物多樣性工作納入各部會年度施政計畫，顯見政府推動之決心。

台灣原生維管束植物約有4,000種，約佔全世界的1.5%，相對於陸地面積所佔的比例，植物種類就顯得非常豐富。為了保護多樣的植物物種，台灣地區已公告共計75處自然保留區、國有林自然保護區、野生動物保護區及國家公園等保護範圍，其中38處由林務局所管轄，顯見林務局在保護生物多樣性及永續性方面，努力不懈。

中華民國企業永續發展協會為因應2001年國際生物多樣性觀察年，特地藉由出版本書作為台灣珍貴及稀有植物紀錄之專輯，讓國內外各界得以瞭解台灣珍貴的植物物種寶藏，並提倡生物多樣性之保育觀念。林務局以全國最大生態資源管理及保育機關之角色，極願意參本書的出版，也希望藉由此書，讓讀者能夠深入認識並省思，為那些即將消失的台灣特有生物尋根，給他們更優越更適合的生存空間。

行政院農業委員會林務局　局長

黃裕星 謹識

Foreword 3

The high degree of attention has been given to biodiversity and nature conservation by most countries of the world, this can be attested to the 180 nations and economy bodies which are signatories to the Biodiversity Convention. In Taiwan, biodiversity conservation has already risen to the level of the Executive Yuan, which in August 2001 passed Resolution 2747, the "Biodiversity Promotion Act ". With various agencies executing, since July 2001, the work of biodiversity has been integrated into the annual plans of various ministers and departments, and it is clear that the government is determined to promote this item.

There are more than 4,000 native vascular plant species in Taiwan. This number accounts for approximately 1.5% of species on earth, and represents a very high proportion to the land area of Taiwan. These figures demonstrate the great richness of plants on the island. In order to protect the diversity of plant species, government has established 75 protection areas, including national forest nature reserves, wildlife protection areas and national parks. It is clear that Taiwan Forest Bureau is doing its part in protecting biodiversity and in sustainability.

To celebrate the Observation Year for Biodiversity 2001, the Business Council for Sustainable Development made the conscious decision to publish this book as a guide to Taiwan's precious and rare plant species. It is hoped that people in Taiwan and abroad will better understand the treasure of Taiwan's precious plant species and become more aware of conserving biodiversity. Taiwan Forestry Bureau is the largest natural conservation and resources management organization in Taiwan, and is very honored and pleased to participate in the publication of this book. We hope that this book will allow the reader to understand and can think deeply on this topic in order to search out the roots of these unique Taiwan plants which are on the verge of extinction, and to provide them with a better, more suitable living space.

Yue-hsing Huang

Yue-Hsing Huang
Director General
Taiwan Forestry Bureau, Council of Agriculture, Executive Yuan, R.O.C

序四

台灣是太平洋邊緣的一顆綠寶石，是世人眼中的美麗之島。曾幾何時，大家埋首在經濟成長之時，早已忘卻這個屬於大家的綠色資產。

山高水急是台灣的重要特色，東部清水斷崖山地直降海岸的壯觀景緻，更是當時被譽為「formosa」的重要場景。台灣有高山、有海洋、有平原、有溪流，各式各樣的環境，孕育了多樣化的生存空間。因為是位在熱帶北緣，所以低海拔有熱帶、亞熱帶的環境；也因為地質年紀輕，山地高低起伏，山頭林立，生活空間多，再加上3千多公尺以上的高山，所以有溫帶、寒帶的環境，而冰河期未受嚴重侵襲的事實，保留了許多第三紀的孑遺生物，更因為雨水充沛，所以生機旺盛，因此台灣實在是一個適合多種生物，包括人類居住的地方，所以孕育了多樣化的生物。

本書的目的在於藉由台灣自然生態系的介紹，讓眾人面對台灣這塊土地上的各種生態環境及一些植物。我們希望藉由此書的出版，在台灣人民已經遺忘了她們的存在之時，喚起大家對於自己腳下這塊土地的一點記憶，重新找回對綠色的、自然的、美麗的台灣的記憶和愛。書中以台灣垂直分布的生態帶為主軸，由低海拔至高海拔地區，包括低海拔地區的海生植被、紅樹林、高位珊瑚礁、熱帶闊葉林、楠木林，中海拔地區的的樟殼林、檜木林，高海拔的鐵杉林、雲杉林、冷杉林、寒原等環境，以及離島地區如彭佳嶼、龜山島、澎湖、小琉球、蘭嶼、綠島及小蘭嶼等，此外尚有比較特殊的維管束植物類群，如東亞孑遺植物—昆欄樹、野生蘭、野生杜鵑花、附生植物、寄生植物、石灰岩區植物、秋海棠、蕨類植物、箭竹、高山野花、菊科植物、龍膽等，分別闡述台灣幾種主要生態系及特色種類。

台灣是北迴歸線上的幸運之島，我們何其有幸生活在這個美麗之島，如果不能好好愛惜這些綠色資產，不僅有愧於此一天造地設的美麗家園，更將為自己帶來嚴重的的浩劫。

台灣大學植物系、所　副教授

郭城孟 謹識

Foreword 4

Taiwan is a verdant island on the edge of the Pacific, an island beautiful in the eyes of man. Not too long ago, everyone was totally engrossed in economic development, and they lost sight of the green treasure belonging to us all.

Taiwan is noted for her high mountains and rushing waters, and the breathtaking landscape of clear waters and precipitous cliffs on Taiwan's east coast was the main reason the island received the label "Formosa" [beautiful] long ago. Taiwan can count mountains, ocean, plains and streams, and virtually every type of natural environment is represented here. The island is replete with diversity. Because of her position on the northern fringe of the tropics, Taiwan has tropical and semi-tropical environments at low altitudes, and also due to her young geological age, her mountain ranges rise and fall and their slopes are covered in forests, providing diverse habitats. Moreover, Taiwan has mountains that tower more than 3,000 meters in altitude, so that she also can count temperate and arctic environments, and the influence of the Ice Age have left behind many traces of Third Age relic species. Because of the pounding rains and plentiful water, life here is abundant, and Taiwan can truly provide suitable habitats for many forms of life, including humans. This is why such a rich diversity of plant and animal life can be seen on the island.

The goal of this book is to use an introduction to Taiwan's natural biomes to bring readers face-to-face with all kinds of living environments and various types of plants. We hope that the publication of this book at a moment when the people of Taiwan have already lost and forgotten the existence of such things will call on everyone to turn their steps toward a memory of this place and once more seek out and return to a memory of and a love for green, natural, beautiful Taiwan. The book is arranged around the natural biomes of Taiwan, moving from the lower altitudes upward. It includes low altitude marine vascular plants, mangroves, forests on higher altitude coral reef, tropical broadleaf forests, and the lowland Machilus forests, the mid-altitude camphor and oak areas, and the high altitude forests of hemlock, Taiwan spruce, Taiwan fir, and the arctic plains. It also includes offshore areas such as Pengchiayu, Kweishan Island, Hsiao Liuchiu, Orchid Island and Green Island, as well as Hsiao Lanyu. Moreover, the book also addresses Taiwan's orchid species groups, such as the relic species, wild orchids, wild rhododendron, epiphytes, parasites, plant species of the limestone cliffs, the begonias, ferns, the short bamboss, alipine wildflowers, chrysanthymums, and the gentians amid descriptions of the special characteristics of Taiwan's major areas.

Taiwan is most fortunate in her location just north of the Tropic of Cancer, and we are indeed fortunate to live on this beautiful island. If we do not carefully treasure our green resources, not only will that cause damage to the beautiful green garden that is Taiwan, but by so doing we will also cause ourselves serious damage.

Chen-Meng Kuo

Chen-Meng Kuo
Curator of the Herbarium,
Associate Professor, Department of Botany,
National Taiwan University

攝影／陳吉鵬

生態空間多樣化

台灣生態環境的特色

Discovering Green Taiwan

◎撰文／郭城孟 ◎ Text ／ Chen-Meng Kuo

攝影／蕨類研究室

■清水斷崖。攝影／蕨類研究室

台灣土地面積狹小，僅36,000平方公里，佔全世界陸地面積不到0.03%，卻擁有相當豐富的生物資源。僅就蕨類植物而言，全世界約有12,000種，台灣就有660多種以上，占5.6%，若以單位面積的種數計算，將近平均值的200倍，是其他地區所望塵莫及的。這全要歸功於台灣擁有極多樣化、可供生物棲息的生態環境，以及未曾遭受嚴重冰河侵襲的事實。也正因為台灣擁有許多不同的棲息環境，冰河期來自不同地區、不同性質的生物才得以在此生存，使得台灣的植物不僅總類多、區系複雜、孑遺種也多。

台灣的環境背景

台灣位在北緯22˚~25˚之間，在這個緯度上的區域多為沙漠或半沙漠的環境，只有在台灣至喜馬拉雅山東部一帶是屬於森林生態系。台灣島上具上百座3,000公尺以上的高山，地形起伏變化很大，造就許多微環境，提供許多不同的生物棲息空間。位在北回歸線上的台灣，是屬於熱帶氣候區的北緣和溫帶氣候區的南緣，因此在低海拔地區形成熱帶及亞熱帶之森林生態體系，而在高山林立的中央脊樑山脈，

Taiwan is small in land area, with just 36,000 km^2. It accounts for less than 0.03% of the world's land mass. However, it possesses a rather rich collection of ecological resources. Among the ferns alone, of the approximately 12,000 species known throughout the world, more than 660 can be found in Taiwan. If we calculated based on unit area, this figure is 200 times the average, and other regions find it hard to come close. This richness is attributable to Taiwan's varied and habitable ecological environment, and the fact that the island has never been subjected to strong attack by glaciers. Precisely because Taiwan possesses many different habitats, the varied animal and plant life that came from many different places during the glacial period was able to settle down here, and this meant that there are many species, many floristic elements and many relic species among Taiwan's plant life.

Taiwan's Environmental Backdrop

Taiwan is located between 22˚~25˚N, at which latitude most areas are desert or semi-arid. Only in the area between Taiwan and the eastern portion of the Himalayan Mountains do the lands at these latitudes belong to the forest biomes. The island of Taiwan has more than 100 mountain peaks of more than 3,000 meters elevation, and there are huge variations in topog-

■連綿不絕的山脈。遠方的山稜即為淡水河的源頭處－聖稜線。　攝影／蕨類研究室

因溫度隨海拔上升而遞減，正好提供了暖溫帶至寒原各類生態系之條件。此造成台灣小而侷限的土地，卻分化孕育出從赤道到北方極地的各種植被形相，使得台灣具有北半球之各種生態環境。

台灣的地質史與區系特色

台灣是位於歐亞大陸邊緣的大陸島，由於距離陸地很近，在第四紀之前浮出水面，其後的數百萬年間，與大陸有幾次陸連，因此歐亞大陸的第三紀古老物種得以進入台灣，其間有四次規模較大的冰河期，台灣都未覆冰，因此植生未受到毀滅式的傷害。大約一萬年前，最後一次冰河結束之後，全球氣溫回暖，逐漸上升的溫度使得適應冷涼的物種無法於原地生存。隨著冰河北退，部份生物向較冷的北方遷移，部份則移往另一較冷的區域，即往高海拔地區遷移，兵分兩路的結果，加上彼此的隔離，逐漸演化出不同的種。因此緯度較低地區的高山和緯度較高的北方會有親源相近的種類。而台灣高山起伏，往高處遷移的物種，分散到各個山頭生存下來，而位於較低海拔溫暖地區的個體則相繼死亡，於是形成不連續分佈的現象，喜馬拉雅山地區和台灣具有類似的種類也是此一原因。因此台灣高山的植物和喜馬拉雅山地區的相似，低海拔的植物則和華南

raphy which create many micro-environments. These provide a habitat for many different species of life. Located on the Tropic of Cancer, Taiwan is on the northern fringes of the tropical areas and the southern edge of the temperate zone, for which reason tropical and subtropical forest is seen at the lower elevations; meanwhile, in the high mountain peaks of the central mountain range, the great reduction in temperatures seen with the rise in altitude provides a range of climates from temperate to arctic. This means that even in the very limited area Taiwan occupies, the island has plant cover from the tropical to the polar, which gives Taiwan a taste of every climate occurring in the whole northern hemisphere.

Taiwan's Geological History and Zonal Characteristics

Taiwan is a large island off the coast of the Eurasian land mass, and because it lies close to the mainland, during the several million years since it poked its head above the surface of the water, it has been connected to the mainland on several occasions. Consequently, ancient life forms from the Tertiary entered Taiwan from the Eurasian mainland, and in the ensuing four major glacial periods, Taiwan was not covered in ice. As a result, plant life was not destroyed on Taiwan. Approximately 10,000 years ago, after the disappearance of the last glacier, the temperature of the Earth warmed and gradually the warmer temperatures made it impossible for species adapted to cold to continue to survive in their original locations. With the withdrawal of the glaciers northward, a portion of the life forms moved to the colder northern regions, and a portion moved to another relatively colder region, the high altitude regions. The results of this bifurcation plus the isolation between the two groups meant they gradually evolved into distinct species. As a result, in the high mountains of lower latitudes and in the far north, there may be similar species. Taiwan's very high mountains were the home of some of these life forms.

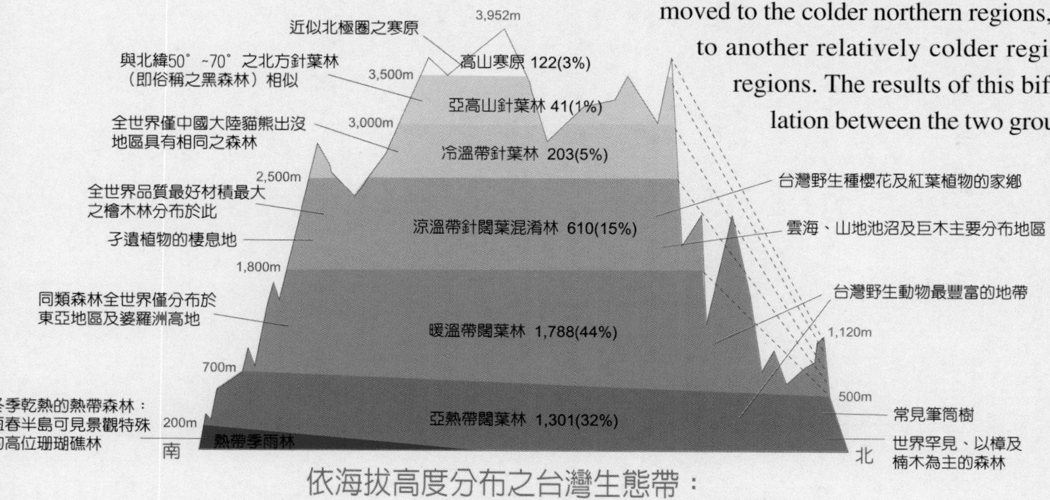

依海拔高度分布之台灣生態帶：
（一）各帶植物種數及所佔比例；（二）台灣各生態帶之重要特色

製圖／郭城孟

較像，且隨著緯度愈高，屬於熱帶的東南亞成份愈來愈少，亞洲成份愈大。

台灣的植物地理親緣關係相當複雜，這是因為台灣位在幾個不同區系的邊界，其地位就像是不同區系的橋樑，是各區系的推移帶，也是生物南來北往的中繼站。北方南來的親潮和南方北流之黑潮，會有許許多多的生物順流至此，而留置島上；北方的大陸氣團、南方海洋氣團也同樣地挾帶著形形色色的動植物來島上落腳。在此南方的熱帶和北方的溫帶交界之處，熱帶植物多分布在南部且向北遞減，而溫帶植物則偏在北部，向南逐漸減少，這是台灣低海拔地區植物生態分化的主要因素。

■ 蘭嶼的熱帶森林，所有樹木都差不多高，形成連續的樹冠層，不會特別高大的突出樹冠層的種類。

攝影／蕨類研究室

台灣生態的全球觀

由於地球繞著太陽運轉有一傾斜角度，因此隨著公轉的軌跡，太陽直射角度會有所差異。太陽直射地面時的溫度最高，隨著緯度的增加，太陽入射角度愈小，地面受到的太陽能愈少，因此愈往兩極的地方，溫度愈低。此外，同一地點的太陽入射角度也會隨時間變化，此一差異每一年循環一次，也就是造成春夏秋冬四季變化的原因。由於在南北回歸線（緯度23.5°）以外的地區，太陽都是斜射的，在極圈（緯度66.5°）以外的地區甚至造成連續24小時的白天或黑夜，此造成隨著緯度之不同，生態環境亦有所不同。一般而言，依緯度分區，可分為回歸線以內的熱帶，極圈的寒帶地區，以及介在兩者之間的溫帶地區，而位在北回歸線上的台灣，是橫跨熱帶及溫帶的環境，因此兼具兩帶的特色，加上台灣山脈高近4,000公尺，因此也出現了與寒帶類似的生態環境。

They distributed themselves among the peaks and settled down to live, and those that stayed behind in the lower elevations died out. This produced discontinuities in the geographical distribution of these species, and is one of the reasons why the Himalayan mountain region and Taiwan have similar species. Consequently, there are similarities between the plant life in Taiwan's high mountains and that of the Himalayas and the plants at lower elevations are more like those in southern China. As the latitude increases, there are less and less S. E. Asiatic elements and the Asian elements are more in evidence.

Taiwan's plants have a complex relationship to the island's topography. This is because Taiwan is situated on the edges of several zones, and she is like a bridge between the various zones, linking them together, as well as a way station for life moving north or south. The Oyashio current, which runs north to south, and the Kuroshio, running south to north, bring many types of life to Taiwan, and some stay on the island; The continental apron in the north and the maritime air mass to the south also bring all kinds of life to the island. At the margin between the southern tropical zone and the northern temperate zone, tropical plants are mostly distributed in the southern regions and move northward, while temperate plants tend to stay in the north, and become scarcer as one moves southward. This is a major factor in the ecological distinctions among plants in Taiwan's low-altitude regions.

A Global View of Taiwan's Ecology

Since the Earth revolves around the Sun on an oblique axis, the angle of the Sun's rays varies at different points on the globe. The times when the temperature is highest, when the Sun's rays hit the Earth directly, as latitude gets higher, the angle of the Sun's rays decreases, and the surface of the Earth receives less energy from the Sun. Consequently, the farther we move toward the two poles of the Earth, the lower the temperature. Moreover, the angle at which the Sun's rays hit the Earth in any given place varies with the time of year. This cycle of change is completed each year, which variation creates the four seasons. Owing to the fact that the Sun's rays always strike at an angle in the areas outside the tropics (latitude 23.5°), and areas beyond the Arctic or Antarctic Circles (latitude 66.5°) may even see 24 hours of daylight or sunless night, there are variations that occur with changes to latitude, and the ecological

熱帶地區

南北緯23.5°以內的地方，也就是南北回歸線之間，是具有太陽直射地面的機會，這個區域稱為熱帶地區，在熱帶地區的生態系包括熱帶雨林、熱帶季雨林、熱帶稀樹草原等。熱帶雨林是終年濕熱的環境，雨林的樹冠層高約30~40公尺，且具有高達50~60公尺突出樹冠層的樹木，例如望天樹等龍腦香科的植物。雨林的草本層和灌木層不發達，這是因為樹木的枝葉相當茂密，林下的光線不足維持光合作用所需，因此地被植物不易生長。雨林植物的葉子大多末端都具有尾尖，方便將積在葉面上的水排出。此外，雨林植物的葉型較大，且常成為裂葉或複葉，這樣可以形成光斑，讓同株植物下方的葉子也可以照到陽光，增加了植物生存的空間。

environment is likewise different. Generally speaking, if we divide the Earth by latitude, we see the tropical region between the Tropic of Cancer and the Tropic of Capricorn, the extremely cold polar areas, and the temperate zone in between these two. And Taiwan, located on the Tropic of Cancer is an environment that includes both the tropical and temperate regions. As a result, Taiwan possesses the special characteristics of two zones, plus Taiwan's mountains, almost 4,000 m in altitude. This is why Taiwan has some ecological environments that are very similar to the arctic environment.

The Tropical Zone

The area within 23.5 degrees North and South latitudes, that is, between the Tropic of Cancer and the Tropic of Capricorn, receives the direct rays of the Sun and is called the tropical zone. The tropical biome includes tropical rain forest, tropical seasonal forest and tropical savannah. Tropic rain forest is hot and wet year-round, with tree canopies of from 30 to 40 meters in height. The emergent canopy trees may reach 50 to 60 meters, for example, in the case of the Dipterocarpaceae plants. The understories are not well developed in the tropical rainforest, because the leaves of the trees are extremely dense and there is insufficient light below the trees to support the demands of photosynthesis. Consequently, the ground cover plants find it difficult to grow. The ends of most leaves on rainforest plants are pointed to facilitate the dispersal of water that collects on the leaves. Moreover, the leaves of these rainforest plants are large, and are often split or compound because sunspots can be formed thus, in order to allow the sunlight to reach the leaves below and increase the living space for the plant.

■南美地區熱帶雨林，樹木高大，隨處可見半人高的板根。 攝影／蕨類研究室

As rainfall varies, different biomes may appear in the tropical region. For example, tropical seasonal-rain forest is a type with dry and wet seasons. The arbor here often is deciduous, and the density of trees in some areas can become very low ; in areas with relatively less rainfall, plants are shorter, and there is no lack of thorny plants. When rainfall is scarce, trees become sparser, and are replaced by shrubs and grasses. This is the savannah. There are many animals in the savannah, such

隨著雨量的變化，熱帶地區還會出現不同的生態系，例如季雨林，這是一種具有乾濕季的森林，其喬木常具有落葉的現象，有些地區的森林喬木密度會變得較稀疏；在雨量較少的地區森林則呈現矮化現象，其中不乏有刺植物。當雨量更稀少時，樹木就會變得更零星，取而代之的是灌木和草本植物，此即稀樹草原。稀樹草原的動物很多，像是長頸鹿、羚羊、獅子等，都是我們所熟知的。

台灣的南部並不是真正的熱帶雨林，因為森林樹冠層偏低，冠層較高的森林多屬於溪谷型的森林，都不具突出樹冠層的種類，

as giraffe, antelope and lions with which we are all familiar.

The southern part of Taiwan is not a true tropical forest, because the forest canopy is fairly low, and the higher canopy forest is mostly located along streams. There are no emergent canopy species, and there is seasonal variation in rainfall, so this area basically is a tropical seasonal-rain forest. However, a number of tropical rainforest phenomena can still be seen in Taiwan, for example, trees with buttressed roots, twining and strangulation, and cauliflorous plants.

The Temperate Zone

■東南亞地區雨林，地面到處積水，需要架棧道才能通行。 攝影／蕨類研究室

The area between the tropics and 60 degrees latitude is the primary location of deciduous forests and prairies. It is also the area that produces the majority of the world's food grains. Apart from prairie and deciduous forest, there are also temperate broadleaf evergreen forests. There are differences on the dominant tree species in temperate forest depending on the area, temperate broadleaf evergreen forest is primarily made up of the Fagaceae, which in Asia are mostly narrow-leaved varieties. In Europe and the Americas these species are mostly wavy-edged; the deciduous forest varieties in Europe are the beech, the leaves of which turn yellow, in America, it is the maple, whose leaves turn red. In Taiwan, both beech and maples occur, and beech forests occur only in northern Taiwan, at the border of Taipei, I-lan, and Taoyuan County on the northern margins of the Snowy Mountain chain, between elevations of 1,500 to 2,000 meters, and the Taiwan red cypress forest belt

■非洲稀樹草原地區，長頸鹿正低頭吃樹上的葉子。 攝影／蕨類研究室

此外，雨量因有季節性的差異，基本上是屬於熱帶季雨林的環境。不過，台灣仍可以看到一些屬於熱帶雨林的現象，例如板根、纏勒及絞殺現象、幹生花等。

溫帶地區

北回歸線以北至北緯 60° 左右是落葉林和草原的主要分布區，也是世界的糧食的主要生產地，除了草原及落葉林之外，也有溫帶常綠闊葉林。闊葉林的種類隨著地區的不同而有差異，常綠闊葉林主要優勢樹種是殼斗科的植物，在亞洲主要是披針形葉的種類，歐美地區的種類則是波浪緣的；落葉林的種類在歐洲是老葉變黃的山毛櫸，在美洲則是紅葉植物一楓樹。台灣則山毛櫸和楓樹都有，山毛櫸林祇出現在台北、宜蘭、桃園縣交接處的雪山北部尾稜，海拔約 1,500~2,000 公尺之間，和檜木林帶差不多同高。而楓樹則範圍較大，約在海拔 1,500~2,500 公尺之間，是演替初級的先峰林。

檜木林則是溫帶地區比較特別的林型，目前地球上的檜木林分布在太平洋兩岸，即東亞東岸的日本和台灣，以及美國西海岸，這些地方都是地質年紀輕、地形陡峭及受海洋性氣候影響容易形成山地雲霧帶，崩塌地形及雲霧是孕育檜木小苗極佳的條件，而台灣的檜木林是全球面積最大的檜木林。

is at about the same altitude. The maple forests are more widely distributed, from approximately 1,500 to 2,500 meters in elevation, and are pioneer forests which replacing the original vegetation.

The red cypress forest is a special feature of the temperate zone. Currently, cypress forests on Earth occur on both sides of the Pacific, in Japan and Taiwan, and on the west coast of the United States. These areas are all geologically young, have a protuberant topography and are influenced by maritime climate, so that mountain rains are common. Landslide topography and fog are excellent conditions for cypress shoots, and Taiwan's cypress forests account for some of the largest mass of cypress forests in the world.

The Arctic Zone

The areas above 60 degrees latitude are the boreal forest and arctic tundra regions. In North America, Siberia and northern Europe, which are close to the southern reaches of the arctic tundra, They are shaped like a belt, and are mostly evergreen conifers, including spruces, firs, and cedars. These trees are similar in shape and height, with straight trunks, conical canopies, equally developed side branches that aim downward, and needle shaped leaves. The forest structure is unistory in tree layer. This type of forest is also called a northern conifer forest. The reason they are called "northern conifer forests" is because this type of forest occurs only in the Northern Hemisphere, since at south latitude 60 degrees and higher except frozen Antarctica, there is only sea water. No forests can develop. This is why the Southern Hemisphere cannot develop this type of conifer forest. Although Taiwan is located in the tropics, because her central mountain range reaches altitudes of 4,000 meters, the high altitude areas have an arctic climactic environment, and so the conifers can grow. Taiwan is the southernmost point for many cold-weather species. In high latitude conifer forests, in the colder north the spruce is found, with fir to their south. Taiwan's conifer forests are fir in the higher elevations, and spruce below them. Moreover, this may be because the fir was a later arrival to Taiwan, and consequently is not distributed in a belt pattern. In any case, the hemlock, with a primary distribution is in Taiwan, Sichuan and Hubei, forms a belt of hemlock forest below the fir. This is the major difference between the conifer forests of Taiwan and the northern conifer forests.

■北美優勝美地（Yosemite）國家公園裏的殼斗科植物，可見其波浪狀的葉緣。

攝影／蕨類研究室

寒帶地區

北緯60°以北的森林是北方針葉林和極地寒原的範圍，在北美、西伯利亞、北歐，緊靠極地寒原的南方，北方針葉林幾乎形成帶狀，其優勢種類多為常綠針葉樹，包括雲杉、冷杉等，它們的形狀和高度相似，樹幹筆直，樹冠錐形、側枝平展至下垂、線形針狀葉，森林結構為單一喬木層。稱為「北方針葉林」的原因是這種森林只出現在北半球，在南緯60°以南除了冰天雪地的南極大陸之外，大都是屬於水域環境，不會有森林的發展，所以南半球不產生類似的針葉林。而台灣雖然位於亞熱帶，但是因為脊樑山脈高達4,000公尺，在高海拔地區具有寒帶的氣候環境，因此也有針葉林的生長，台灣是許多寒冷地區生物最南的分布點。在高緯度地區的針葉林，較冷的北邊是雲杉，冷杉則在雲杉南方，台灣的針葉林則是冷杉在最高海拔，雲杉則佔據在冷杉之下，此外，可能是因為雲杉較晚進入台灣，因此不成帶狀分布，反而是主要分布在台灣及中國的四川、湖北一帶的鐵杉，在冷杉下方形成鐵杉林帶，這是台灣針葉林和北方針葉林極為不同的地方。

北回歸線上的幸運之島

台灣的地理位置相當特殊，是北方溫帶環境的南緣和南方熱帶環境的北限、是北方南下的寒冷洋流和南方北上的溫暖洋流的交會帶、是海洋版塊和大陸版塊的交界。這些不同環境相會的特質，使得台灣

Conclusion

Taiwan's geographical location is very special. Taiwan is on the southern fringe of the temperate zone and the northern edge of the tropical area, between the cold currents coming southward and the warm currents moving northward, and at the border of the continental shelf and the sea. The meeting of the special characteristics of these varied environments means that Taiwan has two faces. If Taiwan lay farther to the north, the tropical environment would disappear, and if she were moved southward, Taiwan would be like Borneo, and would have no conifer forests, no tundra, and of course would never see snow.

■阿拉斯加的寒原及針葉林。

兼具雙邊的色彩。如果台灣的位置往北方偏，則熱帶的環境將消失，往南偏移，則台灣會和婆羅洲一樣，沒有針葉林，沒有寒原，當然也看不到雪。而往東偏則會使台灣成為一個海洋島，而無法成為歐亞大陸冰河期的生物避難所，如此一來，台灣中海拔以上的孑遺植物，將不存在。而若台灣位在比現在偏西一點的位置，則生物會因為和大陸交流頻繁而較少演化出新種的機會。因此，台灣有幸位在此一地區，因其特殊的地理位置，而成為具有多樣化生態空間的美麗之島。

If Taiwan were moved eastward, she would become a marine island, and would not have been able to become a haven for Eurasian species to escape the glaciers; if this were the case, Taiwan's mid- and high-altitude relic species would no longer survive. And if Taiwan were farther to the west, her species would lack their unique space for development due to their interaction with species on the mainland, which would mean that Taiwan will have less opportunity for evolving new species. Taiwan is located in a fortuitous position, and this is what has made her a beautiful island with a greatly diverse ecological environment.

攝影／蕨類研究室

台灣植物多樣性

台灣植物在世界的地位

The World Position of Taiwan's Plants

文／彭鏡毅　◎Text／Ching-I Peng

攝影／彭鏡毅

司馬庫斯的紅檜。攝影／鍾國芳

台灣原生維管束植物約有4,000種，約佔全世界的1.5%，相對於陸地面積所佔的比例，植物種類顯得非常豐富。植物物種單位面積密度與鄰近的菲律賓及日本相比，若以台灣的植物物種密度為1，則日本及菲律賓的比例分別為0.03及0.33，由此可見台灣物種之豐富。在各類群植物方面，台灣蕨類植物有38科、裸子植物有8科及被子植物228科，分別與全世界蕨類植物65科、裸子植物11科及被子植物291科相比，台灣所擁有的各類群植物所佔的比例都超過50%，更可見台灣植物物種歧異度之高。

因台灣位於馬來西亞植物區系與東亞植物區系的交界處，台灣同時擁有許多這兩個植物區系的植物與過渡地區之特殊物

About 4,000 species of vascular plants are indigenous to Taiwan. This number accounts for approximately 1.5% of world plant species, and represents a very high proportion compared to the land area of Taiwan. This demonstrates the great richness of plant life on the island. Compared with the Philippines and Japan, both of which are densely vegetated and close to Taiwan, if we take the density of plant on Taiwan as 1.0, Japan and the Philippines score only 0.03 and 0.33 respectively. This shows us the degree of wealth Taiwan possesses in her plants. In terms of classification, Taiwan is home to 38 families of ferns, 8 of gymnosperms and 228 of angiosperms, and compared to the world total of 65 fern families, 11 of gymnosperms and 291 of angiosperms, Taiwan has more than half of each ranks. This is even more indicative of the high diversity of plant life on Taiwan.

■子遺植物 -- 楓香，樹幹筆直，樹型較像裸子植物。 攝影／蕨類研究室

Taiwan stands at the juncture of the Malaysian and Southeast Asian flora, for which reason Taiwan possess many of the plants of both flora at the same time, as well as special transitional species. Because of her location on the continental shelf of the Asian continent, Taiwan is a key thoroughfare for the southward migration of cold-weather plants from the Eurasian landmass and the northward movement of tropical plants. Many plant species moved to Taiwan's middle-altitude mountains during the Ice Age, and became relic species there. These include Taiwan red cypress, Japan cypress, *Taiwania cryptomerioides* Hayata, *Dysosma pleiantha* (Hance) Woodson, Trochodendraceae, Chinese sweetgum (*Liquidambar formosana* Hance), and Taiwan andromeda. These species are all relic species originating in the Northern Hemisphere. They further explain the important role Taiwan plays in plant evolution and historical migration, and the climate changes to the East Asian can be thoroughly investigated on the basis of these species and the geographical characteristics they prefer. They also contribute data to our knowl-

種，且由於位在亞洲大陸棚東緣，為北方歐亞大陸寒帶植物南徙與南方熱帶植物北遷的交通要道，許多植物在冰河時期遷徙至台灣，間冰期退縮至中高海拔山區，成為冰河孑遺植物，例如紅檜、扁柏、台灣杉、八角蓮、昆欄樹、楓香、台灣馬醉木等，這些都是北半球原始的孑遺植物，也進一步說明台灣植物在物種演化及歷史變遷上所扮演的重要角色。藉由不同物種的植物分類與親緣地理學研究可以探知東亞地區氣候的變化，與植物遷徙及演化的歷史。台灣的高海拔植物物種豐富，具有北方大陸各種代表性植物生態系。由於地質時代氣候變遷的影響，台灣中海拔 1,000~2,000 公尺的區域有著許多古老植物的成分；北部低海拔地區主要屬於亞熱帶森林生態體系，南部地區則為熱帶的北界，更因黑潮流過台灣兩側，使台灣氣候比同緯度地區溫暖而終年蓊鬱。

除了特殊的地理位置成就了台灣植物在世界上的特殊地位之外，台灣地質年代甚輕，由於受板塊運動的影響，山勢高低起伏，提供了細膩多變的生物棲息空間，並且隨著海拔上升，豐富了各類生態系形成之條件，孕育出許多不同的生態環境，

■楓香的花。 攝影／彭鏡毅

edge of plant evolution and migration. Taiwan's high-altitude regions are abundant with plant species. This area has all the representative plant ecosystems of the northern mainland, and because of the impact of climate change over the geological ages, the mid-altitude region from 1,000 to 2,000 meters have many ancient plant species. The low-altitude regions in the north are largely sub-tropical forest ecosystems, while the south mostly shows tropical conditions. Because the Kuroshio current passes by both sides of Taiwan, Taiwan's climate is more temperate than other countries at the same latitude, and is luxuriant the year round.

■楓香的果。 攝影／彭鏡毅

■馬醉木是北半球的孑遺植物之一。

攝影／鍾國芳

而成為北半球各類生態系的縮影，加上海洋與高山的隔離，得以演化出許多獨特的物種。在台灣4,000種原生植物之中約有1,000種植物為台灣特有種，其比例約佔四分之一，在東亞首屈一指，雖然不若夏威夷（89%）、紐西蘭（85%）等海洋性島嶼，且台灣亦無特有的科，但在東亞植物區系裡，台灣植物在植物地理及演化上佔有一個不容輕忽的重要關鍵地位。

台灣依據地形與氣候大致可分為東北部地區、東部區、北部區、西南部區、中央山脈區及南部區。台灣東北部地區首當東北季風之衝，終年潮濕多雨，東部地區主要在夏秋兩季受颱風影響，雨量豐沛，北部地區多丘陵台地，西南部地區為廣闊之平原，而中央山脈為台灣的脊樑，隨著海拔提升氣候從溫暖轉為酷寒，而南部地區的物種歧異度獨冠全台，可謂各具特色，也因此使得台灣這個小島有著各種不同的氣候地理區，孕育著多樣的植物種類。

台灣地區目前在珍貴稀有植物資源保育方面已有相當成效，為了保護多樣的植物物種，近年來相繼公告共計88處自然保留區、國有林自然保護區、野生動物保護區及國家公園等，總面積約佔台灣陸域面積19.5%；其中並包含針對特定植物物種所劃定的保護區的保留區，例如淡水河紅樹林自然保留區、坪林台灣油杉自然保留區、台灣一葉蘭自然保留區、台東蘇鐵自然保留區、台灣穗花杉自然保育區等，足見台灣地區內珍貴稀有植物之多樣性與重要性。如此大面積的保護對於植物的保育是最直接有效的方式，除能提供直接的保護之外，更是資源保育研究的基礎。

Apart from her special geographical location, which gives Taiwan's plants a special position in the world, the influence of tectonics, Taiwan's young geological age, and the island's mountain ranges have provided a wide variety of habitats for plants. Moreover, as altitude increases, the conditions are created for nearly every type of ecosystem on earth. This gives birth to many different environments, and creates a microcosm of the northern hemisphere. Furthermore, Taiwan's high mountains are not far removed from the sea, meaning that many unique plant species have evolved. Among the 4,000 species of plants native to Taiwan, approximately 1,000, or one-fourth, are unique to the island, putting Taiwan in first place among the Southeast Asia/China region. Although she cannot compare with Hawaii (89%) and New Zealand (85%), which are oceanic islands, and although Taiwan has no plant families unique to this island, in the Southeast Asian flora, the plants of Taiwan occupy a key position in terms of phytogeography and evolution.

Taiwan can be roughly divided into the northeast, east, north, southwest, central mountain region and western regions, based on climate and topography. The east region is hit by northeast monsoons, and feels their effect year-round. The east is hit by typhoons primarily in the summer and fall, and sees ample rainfall. The northern region is hilly, and the southwest consists of broad, flat plains. The central mountain region is the backbone of Taiwan, and the climate here becomes colder as the altitude increases. The southern region is the champion for unique plant species, and it can be said that each has a special characteristics. Consequently, the tiny island of Taiwan actually has almost every climactic area possible, and is the home of an extremely diverse set of plants.

Right now, Taiwan has been quite successful in the preservation of her rare plant species. In order to protect the diversity of plant species, in recent years 88 natural preserve areas, national forest natural preserves, wild animal preserves and national parks have been declared as protected areas, making up a total of 19.5% of Taiwan's land area; among these are some preserves aimed at saving particular species of plant, for example, the Tamsui River Mangrove Forest Natural Preserve Area, the Pinglin Taiwan keteleeria fortunei Nature Preserve, the Taiwan Pleione Natural Preserve Area, the Taitung *Cycas* revoluta Nature Preserve, and the Taiwan *Amentotaxus* Preserve. These areas amply display the diversity and importance of Taiwan's rare plant species, and the protection of large areas of these plants is the most directly effective method to undertake. It not only provides direct protection, but also acts as a foundation for studies of resource conservation.

多彩多姿的世界

低海拔
The Low Altitudes

◎撰文／郭城孟 ◎ Text ／ Chen-Meng Kuo

攝影／蕨類研究室

內雙溪。 攝影／蕨類研究室

■亞熱帶闊葉林冠層枝葉茂密，灌木和地被密佈，中間還穿插了許多蔓藤和著生植物，結構複雜。 攝影／蕨類研究室

台灣從海平面至500公尺（北部）或200~700公尺（南部）以下，是亞熱帶闊葉林。闊葉林的森林結構常見分成三至四層，即一至二層喬木層、灌木層、草本層，各層之間還可以看到蔓藤和著生植物，主要組成植物都是闊葉類植物。由於台灣南部位在熱帶邊緣，因此台灣南部低海拔可見一些熱帶森林生態系。台灣低海拔的闊葉林帶，隨著緯度、雨量等的差異，可以分化出多種不同的生態系，例如屬於熱帶的海岸林、沙灘植群、岩岸植群、紅樹林、高位珊瑚礁林、熱帶季雨林、熱帶雨林等，以及屬於亞熱帶的樟楠林。

台灣低海拔地區最典型的是以樟樹和楠木為主的亞熱帶闊葉林，因此也稱為「樟楠林」，除了樟樹和香楠之外，主要伴生的種類有烏心石、杜英、樹杞、九芎、島榕等，灌木層則以九節木、台灣山桂花等最普遍。但是低海拔是人類活動頻繁的地區，所以大部份自然度較高的成熟環境都已被破壞殆盡，其中樟樹因經濟價值高，而被砍伐殆盡，由香楠取代其地位之外，很多地區因過去柴薪的需求而成為相思樹造林地，還有許多地區則因受到火災或開發的關係，成為五節芒草原。這些地方如果不再受到干擾，會逐漸演替成以野桐、白匏子、血桐、山黃麻等，以陽性植物為主的先鋒林，之後逐漸形成較成熟的、以香楠為主的亞熱帶闊葉林。在溪谷地區則會出現較多的榕類植物，例如水同木、幹花榕、菲律賓榕等，此外，大葉楠也是優勢樹種，其他還有茄苳、長梗紫麻、水冬瓜等，共同組成「亞熱帶溪谷森林」。這類森林如果受到干擾，其先峰林樹種主要是筆筒樹。此外，在台灣北部地區，因東北季風帶來的強風、低溫和多雨，使得

■台灣樹參。 攝影／蕨類研究室

■水冬哥。 攝影／蕨類研究室

The area between sea level and 500 m (in the north) or 200~700 m (in the south) is the subtropical broadleaf forest area of Taiwan. The broadleaf forest is frequently made up of three or four layers, with one or two layers of trees, one layer of shrubs and grasses for each. Vines and epiphytes can be seen, and the primary plant species are broadleaf. Since the southern portion of Taiwan is on the border of the tropics, a number of plants from the tropical forest environment can be seen there. The broadleaf forest in Taiwan's low altitudes varies with the latitude and amount of rainfall. It can be classified into many different ecosystems of the tropical belt, such as the coastal forest, the beach plants, the seaside cliff plants, the mangrove forests, the coral atoll, the tropical seasonal-rain forest, and the tropical rain forest. There is also subtropical camphor forests in the north.

The most typical plants in the low altitude areas are primarily subtropical broadleaf trees such as *Cinnamomum camphora* (L.) Nees & Eberm. and *Machilus zuihoensis* Hayata. Because of this, these forests are also called camphor forests. Apart from these two species, *Michelia compressa* (Maxim.) Sargent, *Elaeocarpus sylvestris* (Lour.) Poir, *Ardisia sieboldii* Miq., *Lagerstroemia subcostata* Koehne, *Ficus virgata* Reinw. ex Blume and others are often seen. In the shrub layer, *Psychotria rubra* (Lour.) Poir. and *Maesa tenera* Mez are the most common species. However, the low altitudes are areas of frequent human activity, so the majority of the highly mature environment has already been destroyed. Moreover, because of the high economic value of camphor, these trees have largely been cut down. Camphor's place has been taken by *Machilus zuihoenis*, and in many areas, the need for firewood in the past has caused a transition to *Acacia confusa* forests. There are also many areas in which fire or development gave rise to *Miscanthus* tall grassland. If these areas are not disturbed again, a gradual evolution will see them replaced by *Mallotus japonicus*(Thunb.) Muell. -Arg., *Mallotus paniculatus* (Lam.) Muell. -Arg. *Macaranga tanarius* (L.) Muell.-Arg, and *Trema orientalis* (L.) Blume as pioneer forest; afterwards there will be a gradual transition to a more mature subtropical broadleaf forest composed primarily of *Michilus zuihoenisis*. More fig tree species appear in the region of streams, such as *Ficus fistulosa* Reinw. ex Blume, *Ficus variegata* Blume var. *garciae* (Elmer) Corner, *Ficus ampelas* Burm. f. and so on. Moreover, *Machilus japonica*

亞熱帶闊葉林。 攝影／蕨類研究室

北台灣地區具有一些比較特殊的種類，例如野鴉椿，以及屬於較高海拔地區的的植物種類，如台灣樹蔘、昆欄樹、台灣石楠、台灣馬醉木等，這種森林稱為「東北季風林」，東北季風林因受強風吹襲，植物體的枝條常被吹折，而形成植株矮化且分枝多、節間短的樹木。

在南部地區以及海岸環境，則出現熱帶的環境，所謂「熱帶」是指南北回歸線間之區域，其中的熱帶森林其特色是年降雨量多且氣候酷熱，但是台灣因為位在熱帶邊緣，所以這些「熱帶環境」都不太典型。蘭嶼是台灣「熱帶雨林」分布的地點，在此可以看到熱帶林的生態現象，例如板根、纏勒、絞殺、幹生花等，但是沒有特別高大的植株，也無突出樹冠層之樹木。

在台灣南部地區，冬季雨量偏少，呈現明顯的乾濕季，植物生長的春季常欠缺水分，而夏季則有颱風帶來暴漲的降雨量，因此台灣南部低海拔地區雖然年雨量、溫度都達到雨林的標準，卻因為年雨量不平均，有季節性的乾旱，因此無形成標準的雨林，故其樹冠層較低矮，沒有突出樹冠層之大樹，且具有落葉樹種，這樣的森林稱為「季雨林」或「季風林」。全世界除了台灣南部以外，中南半島亦有熱帶季風林的分布，柚木是熱帶季風林的代表樹種。

在南部海岸偏內陸地區，是隆起的珊瑚礁，因土壤少、保水差，且含有較多的碳酸鈣，因此所生長的植物比較特別。在恆春半島的社頂、墾丁遊樂區等地可見，主要優勢植物有白榕、毛柿、象牙樹等。由於位居較內陸地區，比海岸林受風壓較小，且無鹽沫的傷害，植物生長較佳，較

Sieb. & Zucc. var. *kusanoi* (Hayata) Liao is another dominant species. Additionally, *Bischofia javanica* Blume, *Oreocnide pedunculata* (Shirai) Masamune, *Saurauja oldhamii* Hemsl. and others make up the subtropical gorge forests. If this type of forest is disturbed, the succeeding pioneer forest is primarily made up of *Cyathea lepifera* (J. Sm.) Copel.. Moreover, in the northern regions of Taiwan, due to the strength of the north-east monsoon winds, low temperatures and the frequent rains, some fairly unique species exist. For example, we can see *Euscaphis japonica* (Thunb.) Kanitz, and plant species normally seen in the high-altitude regions, for example, *Dendropanax pellucidopunctata* (Hayata) Kanehira ex Kanehira & Hatsusima, *Trochodendron aralioides* Sieb. and Zucc., *Pourthiaea lucida* Decaisne, and *Pieris taiwanensis* Hayata. These forests are called the northeast monsoon forests. Since these forests are exposed to the effects of rain and wind, the branches of the plants here are often sheared off, and the trees are usually short with multiple stems and short internode.

In the southern regions of the island, and on the coasts, the climate is tropical. The so-called "tropical belt" of the world is the area between the Tropic of Cancer and the Tropic of Capricorn, and the special characteristic here is a year-round hot and rainy climate for tropical forest area. However, due to Taiwan's location at the fringes of the tropical zone, her "tropical environments" are really not very typical. The island of Lanyu represents Taiwan's tropical rain forests, and here we can see typical features of tropical plants such as buttressed roots, lianas, stranglers, and cauliflorous plants. But there are no particularly emergent tall trees.

In the southern regions of Taiwan, there is little rainfall during the winter season, and there is a marked dry season. The plants here often lack water during the spring growth season, and summer typhoons bring violent storms and large volumes of rain. Consequently, although the southern regions of Taiwan have annual rainfall and temperatures that fit the characteristics of a rainforest, because of the unequal distribution of the precipitation of the year and the seasonal droughts, there is no standard rainforest here. The trees are shorter, and they do not protrude qiant trees through the canopy. Some drop their leaves. This type of forest is called the "seasonal-rain forest". Apart from the southern part of Taiwan, Indo-China is an other place with seasonal-rain forests where teak tree is their represeutative, Peninsula.

In the southern coastal areas bordering on the inlands, there are raised coral reefs. Due to the small amount of soil and poor water retention characteristics here, added to the high amount

潮濕處還可以看到雨林的特徵，如板根現象，此混合了雨林、石灰岩植群、海岸林特色的植物社會，稱為「高位珊瑚礁林」。

在接近海岸線的植物群落，包括臨海珊瑚礁植物群落、沙灘草本植物群落、海岸灌叢以及海岸林。目前完整的「海岸林」僅見於墾丁國家公園的香蕉灣一帶，海岸林的發展主要受限於地形條件，通常發育在距海不遠、避風良好且腹地廣大的地方，然而這些地方也經常是人類喜愛開墾的場所，現在僅零星可見黃槿、海檬果等種類。台灣東海岸主要為陡峭的岩岸，並無廣大腹地可供海濱植物群落發展，而西海岸則多沙灘，主要是匍匐性成叢生長的草本植物，如馬鞍藤、天蓬草舅、濱刺麥、濱當歸、茵陳蒿等所形成的「沙灘植物群落」，以及林投或蔓荊、草海桐、雀梅藤、毛苦蔘等形成之「海岸灌叢」。南部地區因氣溫較高，珊瑚生長旺盛，在珊瑚礁縫中可見水芫花、山豬枷等形成的「珊瑚礁植群」。

而在河口地區因水流速減緩，造成淤沙量大，且海水容易進入，是鹹水和淡水交會的地帶，隨著潮起潮落，水中含鹽度會有差異，這樣的環境下，孕育了特殊的河口濕地生態系一紅樹林。台灣地區的種類包括紅樹

of calcium carbonate, the plants here grow in a rather unique way. On the Hengchun Peninsula at Sheting, Kenting Recreation Area and other areas, we see primarily *Ficus benjamina* L., *Diospyros discolor* Willd., and *Diospyros ferrea* (Willd.) Bakhuizen. Since these areas are more inland, they receive less wind than the coastal zones, and are not damaged by salt water. The plants here grow better, and more typical rainforest characteristics are visible than in the tidal areas, for example, buttress roots. This area is a mixture of rain forest, limestone plants, and coastal forest, and is called the high coral forest.

The plants near the coast include coral reef plants near the sea, the beach grasses, the coastal shrubland and the coastal forest. Currently, climax coastal forest is seen only at Hsiangchiao Bay in Kenting National Park, since its development is primarily limited by topographical factors. Often seen near the sea, in broad areas that are well protected from the wind, this type of areas that are also preferred by humans. Here we can see sporadic examples of *Hibiscus tiliaceus* L., *Cerbera manghas* L. and other plants. Taiwan's east coast is primarily steep cliffs, and there is no broad area of flat land to provide a habitat for the development of coastal forests. The west coast has many beaches, which are the home to primarily prostrate grasses, such as *Ipomoea pes-caprae* (L.), Sweet subsp., *brasiliensis* (L.) Ooststl., *Wedelia prostrata* (Hook. and Arn.) Hemsl, *Spinifex littoreus* (Burm. F.) Merr., *Angelica hirsutiflora* Liu., Chao

■墾丁國家公園的香蕉海岸林。

攝影／蕨類研究室

科的水筆仔、五梨跤、紅海欖，使君子科的欖李，馬鞭草科的海茄苳，以及在台灣已經消失的紅樹科成員：紅茄苳、細蕊紅樹。紅樹林也是熱帶的生態系的一環，在台灣的分布是往北種類逐漸減少。由於紅樹林生長在水陸交接的環境，它的根系可以攔截並累積泥沙，有防波護岸的功能，而這種環境鮮少人工建築能穩定地存在，因此紅樹林是極優良的防潮堤。紅樹林會因洪水而被沖去部份個體，而在新形成的裸地中，會先長出茳茳鹹草和蘆葦等高莖草類，之後才逐漸恢復成紅樹林，在國外有些地方是則以鹵蕨為先峰種類。

但是在一些坡陡流急的小溪，其出海口因水流速較快，淡水較多，且無大面積淤沙，所以沒有紅樹林的分布，取而代之的是以穗花棋盤腳為優勢樹種的溪口林。溪口林主要分布在北台灣地區，恆春半島亦有，其他如淡水溼地也有穗花棋盤腳的分布，例如蘭陽平原和以前的台北盆地等。

濕地是水陸交接的環境，除了前述的紅樹林是屬於河口濕地，在內陸地區還有下游河邊的河岸濕地，以及一些在比較低窪地區的平野濕地。濕地是野生生物覓食棲息的最佳場所，尤其是河岸及河口濕地，由上游沖刷下來的有機物質會隨著河

■水茄冬（穗花棋盤腳）長而下垂的花序。

攝影／蕨類研究室

and Chuang, *Artemisia capillaries* Thunb. and other beach plants. Here we also see *Pandanus odoratissimus* L. f. or *Vitex rotundifolia* L. f., *Scaevola taccada* (Gaertner) Roxb., *Sageretia thea* (Osbeck) M. C. Johnst. *Sophora tomentosa* L. and other beach shrubs. In the south of Taiwan, due to the higher temperatures, coral grows luxuriantly, and at its seams we can see *Pemphis acidula* J. R. & G. Forst, and *Ficus tinctoria* Forst. f., which make up the coral reef plant group.

In the estuarine areas, because of the low speed of water, a great deal of sand is deposited, and seawater can easily flow in. This is the area where the saltwater and freshwater meet, and with the rise and fall of the tides, the saline content of the water changes from time to time. In this environment, we find the special estuarine wetlands -- the mangrove. Taiwan's species include *Kandelia candel* (L.) Druce, *Rhizophora mucronata* Lam., *R. stylosa* Griffich, *Lumnitze raracemosa* Willd., *Avicennia marina* (Forsk.) Vierh., and some mangrove species that have already disappeared from Taiwan: *Bruguiera gymnorrhiza* (L.) Lam. and *Ceriops tagal* (Perr.) C. B. Robins. Mangrove forests are one link in the tropical biome, and they become scarcer in Taiwan as one moves northward. Since mangroves grow in the area between land and water, their root systems can hold and accumulate sand, and they are useful for protecting coastlines. This type of environment has no stable area for people to build on, for which reason the mangrove forests are good for use as protective levees. The mangrove forests may be largely washed out in a flood, and on the newly cleared land, we first see cyperaceous plants and other high-stemmed grasses. These are followed gradually by the reappearance of the mangroves. Outside of Taiwan there are some areas in which *Acrostichum aureum* L. is the pioneer species.

However, in some sloping creeks with fast water flows, sea outlets have such a fast flow of water that the fresh water predominates. There is little sand accumulation here, so there are no mangroves. Replacing the mangroves are *Barringtonia racemosa* (L.) Blume ex DC., the predominant species in the mouths of the creeks. The forests at the mouths of creeks are primarily found in northern Taiwan, but there are also some on the Hengchun Peninsula. Other areas, such as freshwater marshes, also have a few such forests, for example, the Lanyang plain and the former Taipei basin.

The marshes are the environment where water meets land, and apart from the mangrove forests already described above, which are estuarine, inland areas also see riverbank marshes on the lower reaches of rivers. There are some lowland plain areas which have open marshes as well. Marshes are an excellent

■脈耳草最愛海岸的環境。 攝影／桂曉芬

曲、泥灘地的形成而累積在岸邊，許多植物及魚、蝦、貝類在此環境生長，有了棲息空間及食物，於是吸引次級消費者的較大型動物徜徉其間。河川濕地除了是生物的棲所之外，也是水陸生態系的交會帶，它的價值不僅是生態上的，它寬廣的空間可於驟雨或洪氾期調節水量，除了減少洪水出現的頻度，以及降低洪氾的程度，也可減緩洪水流速並防止河岸遭受沖蝕，是洪泛期的緩衝區。然而因為人類與河川爭地，把平常未有水流的河川地當作荒廢地，有的填土種菜，有的用來堆置廢棄物，河川兩岸的濕地因而陸續消失，不但影響野生動植物的生存，也因影響河川行水的流暢性，而危害到人類的安全。與河川爭地的下場通常是在大雨時受到自然的反擊，由於人類佔用行水區，大雨引發的洪水即有可能在原有濕地範圍竄流。

habitat for wild animals to feed, particularly the marshes on rivers and along the mouths of rivers. Because of the organic matter washed down the rivers, which accumulates at the bends or muddy areas of the river, forming banks, many plants and fish, shrimp and shellfish species live in this environment, which provides them with habitat and food. Secondary consumers are also attracted here; many of these are larger animals. The marshes along rivers not only provide excellent habitat, they also are a transitional area between water and land systems which is benifit for ecological function. Their value is not only confined to providing a transition between the two; their broad spaces also collect rainwater and prevent flooding by helping to regulate the amount of water. Apart from reducing the frequency of flooding, and reducing the severity of any floods that do occur, these marshes can also reduce the speed of flow for flood waters and protect the riverbank from erosion. They act as a buffer zone during the flood season. Due to human activities that fight the rivers for land, dry riverbeds are often taken as wastelands, and some are planted with vegetable crops. Some are used to store trash, and in other rivers the marshy areas on both shores have disappeared. This not only impacts the wild animal habitat, but also affects the smoothness of flow of the river, and endangers humans. Places downstream are often affected when heavy rainfall hits. Since humans are occupying the places where water should flow, the heavy rains bring flooding which is dispersed in the areas where humans are, those that were originally marsh.

■低海拔池沼，圖中點點黃色是台灣貍藻的花。 攝影／蕨類研究室

水中舞者

海生植被

Marine Vegetation

◎撰文／楊遠波

◎ Text ／ Yuen-Po Yang

台灣的海岸地區是一個多彩多姿的環境，衆所熟知的海邊生物像是美麗的珊瑚、熱帶魚，以及各式各樣的海藻，都是生活在這個範圍。其中有一群是屬於開花植物，因爲它們開花、授粉都在水中進行，因此花不明顯，花的色彩也不炫麗，長得不太起眼，加上因爲生活在水中，植物體柔軟，隨著浪潮擺動，外形與藻類酷似，故而不是常常被人們遺忘，就是乾脆被當成是藻類，所以認識它們的人可能不多。這些會開花的水生植物被統稱爲「海草」，與藻類的「海藻」不同。海生開花植物，需要陽光進行光合作用，所以通常生活在淺水域，最深也不過四、五十公尺。

Taiwan's coastal zones are a diverse environment. Seaside life that everyone is familiar with, such as coral and tropical fish, rub shoulders with various types of seaweed, but all have this habitat in common. A few are flowering plants, and because they flower and are pollinated in the water, these flowers are unlike most flowers in their muted colors, and do not much attract the eye. Added to this, because they live in the water, these plants have soft bodies, and ripple in sympathy with the tide, with an outer appearance like seaweed. As a result, if they are not often forgotten once seen, then they are dismissed as being nothing more than seaweed. Therefore, few people are able to recognize them. These flowering marine plants are collectively called "sea grasses", but are distinguished from the "seaweeds". Flowering marine plants require sunlight to carry out photosynthesis, so they often live in shallow areas, no deeper than forty or fifty meters.

■泰萊藻，與卵葉鹽藻在台灣的分布相似。

攝影／楊遠波

熱帶到溫帶地區，平緩海岸的潮間帶和低潮線以下地方會有高等的開花植物（flowering plants）生長，形成了海草植被。這些植被由於開花結果以及許多浮游生物和海藻生活其間，提供一群棲身在該群落中和附近環境中各類動物的食物。因為海草的生長，可以穩定水流和沉澱懸浮在水中的小粒子，此類植被也成了那些動物絕佳的棲息場所。

海生開花植物（以後簡稱海草，英名seagrasses）可說是高等植物中甚為特殊的一群，此是因為生長的環境和長在陸地上的高等植物截然不同。通常所見的陸生植物可從生長的環境中取得水分子，而海草卻要從鹹水中取得水分子。此外，這些海草在水中開花和授粉，是為狹義水生植物的成員。

一般人只知有藻類長在海水中，並不知道高等的開花植物也可在海水中生長。其實，不少的植物學家也一樣不知道有這些植物的存在。因此，人們以往對於海草所知不多。到了近一、二十年來，它們的外形和生態上的重要性才漸為人們研究和發掘。

生物學家已瞭解，類似的自然環境會選擇出外形相近但沒有親緣關係的種類。例如美洲和非洲沙漠，兩者的環境相似，而篩選出具有針狀葉和肉質莖的仙人掌科和大戟科植物。海水中的環境如同沙漠氣候一般，因而海草的許多種類外形是那麼的像。大多數的海草有長而具韌性的根莖（rhizome），非常堅固地著生在基床（substrate）上，每節上都長出長條形的葉且沒有葉片和葉柄的分別。

台灣地處熱帶與亞熱帶之

■綠島東北角海水中的泰萊藻群落。 攝影／楊遠波

間，海草自當分布。僅就本島來說，海草大多分布在西海岸，甚少在東海岸。這可能和西海岸平坦及東海岸陡峭有關。西海岸北從新竹香山海濱一帶，往南至台中大肚溪口、彰化新寶、嘉義布袋、屏東的海口、後壁湖、後灣至南灣為止。東部僅在台東成功以北至小港為止。至於離島，所知有海草分布的地方有綠島、小琉球和澎湖。日據時代，曾有北達近台北石門一帶以及南部高雄一帶的海草採集記錄，但迄今這兩地不曾被人見過海草的蹤跡。

世界上所知的海草種類約有五十種，皆屬單子葉植物的眼子菜科（廣義）和水鱉科，生長在各大洋（北極和南極除外）的海岸海水中。台灣至今所知至少有7種，為貝氏鹽藻（*Halophila beccari* Aschers.）、毛葉鹽藻（*Halophila decipiens* Ostenfeld）、卵葉鹽藻（*Halophila ovalis* (R.BR.) Hook.f.）、泰萊藻（*Thalassia hemprichii* (Ehrenb.) Aschers.）、單脈二藥藻（*Halodule uninervis* (Forsk.) Aschers.）、線葉二藥藻（*Halodule pinifolia* (Miki) Hartog）和甘藻（*Zostera japonica* Aschers. & Graebner）。其中的貝氏鹽藻呈零星散佈，不成所謂的群落，一度被認為已在台灣絕跡。其它的六種有小至很小面積的，大至呈數十平方公尺的群落存在。

毛葉鹽藻是小至一、二叢存在於屏東南灣北邊角落近核能電廠入水口一帶的沙質地上。該地水深五至十餘公尺處可見散生的毛葉鹽藻。

南灣的南邊角落是一小面積平坦的珊瑚礁岩砂灘，漲潮時水深及膝，退潮時水僅淹至腳踝。在砂灘上有一片主要由泰萊藻形成的海草群落，中間雜以單脈二藥藻混生。這片是全台灣最容易到達可見的海草植被。

除了南灣外，後壁湖、萬里桐和綠島東北方海邊都有大片的泰萊藻群落。小面積的在小琉球、屏東的海口可見。

■毛葉鹽藻。 攝影／楊遠波

■卵葉鹽藻，在台灣分布廣，可在台灣東部、南部、小琉球、澎湖的岸邊海水中生長。 攝影／楊遠波

■澎湖講美村的海灘，當退潮後，卵葉鹽藻和甘藻形成的群落漸露出水面。 攝影／楊遠波

■ 綠島南寮的海水中有單脈二藥藻、卵葉鹽藻和泰萊藻形成的海生植物群落。

攝影／楊遠波

■ 屏東後灣海灘，海水中有角果藻科之單脈二藥藻等。 攝影／楊遠波

單脈二藥藻除了與泰萊藻混生外，有時也可見小面積的純群落，如在台東成功以北的小港海水中。在澎湖，它也和卵葉鹽藻混生，但不是優勢的種類。

在屏東海口沿海岸邊南行約一公里處，有一片大沙灘。這裡漲潮和退潮時海水深度的落差不大。在沙灘上長了一大片線葉二藥藻。由線葉二藥藻的生長基質來看，似乎和單脈二藥藻的珊瑚礁岩砂地不同。有些學者曾質疑單脈二藥藻和線葉二藥藻為不同種的真實性。從兩者對生長基質需求的不同上或可提供一解決的線索。

甘藻是溫帶性的種類，從海南島可分布達日本。新竹香山海邊有一大片甘藻群落，向南達大肚溪口、彰化的新寶、嘉義的布袋到屏東的海口。大肚溪口、新寶和布袋都是純群落。尤其是布袋，前二十多年在海水中密生。但是自從布袋港及其海岸經整建後似乎已消失殆盡。至於海口海邊則是和卵葉鹽藻混生，似乎仍以甘藻為主。

卵葉鹽藻是海草中在世界上分布最廣的種類。無疑的，在台灣也有很廣的分布範圍。可從屏東的南灣向北分布到海口，在台東由成功北至小港，而後在澎湖成大面積的生長。總而言之，澎湖生長的海草植被可說是集台灣海草的大成，不但面積大，種類也多。

除了上述的地方外，若把東沙島加入高雄縣內，又可多出兩種海生植被，那就是針葉藻（*Syringodium isoetifolium*）和絲粉藻（*Cymodocea rotundata*），這兩種是長在東沙島的潟湖與海岸海水中。

■ 單脈二藥藻在屏東縣後壁湖形成的群落。 攝影／楊遠波

海上森林

紅樹林

Mangroves

◎撰文／薛美莉
◎ Text ／ Mai-Li Hsueh

位在河海交界的紅樹林，是具有生命的天然防波堤，在台灣西部海岸地區，多有它們的蹤跡。紅樹林的生長除了要有熱帶氣候條件外，還必須符合具遮蔽性海岸線、廣大潮間帶及淺灘、維持正常進退的潮汐水流以及富含有機質的泥灘地等環境特性，才能讓紅樹林植物形成大面積群落。日治時期還曾經將基隆灣與高雄灣的紅樹林指定爲「天然紀念物」。台灣的紅樹林植物包括4科7屬7種的眞正紅樹林樹種，以及一些也會經常在此類環境出現的海岸林植物種類。其獨特的胎生苗現象，以及對潮汐、鹹水等的高度耐受性，更是它們吸引大家注意的焦點。

The mangroves, living as they do at the border between the river and the sea, are a natural dyke, and in the western coastal regions of Taiwan, they are often seen. To grow, mangrove forests require not only a tropical climate, but also a sheltering coastline, a broad tidal band and shallow beach, tides that maintain a regular rhythm, and sediment rich in organic nutrients. Only in such conditions will mangroves multiply into a forest and cover a large area. During the period of Japanese rule, Keelung and Kaohsiung Bays were set aside as "Natural Memorials." Taiwan's mangrove forests include plants from 4 families, 7 genuses and 7 species of true mangrove trees, as well as a number of species of coastal plants often found in similar environments. The unique phenomenon of the live birth shoots in the mangrove forests as well as their high degree of resistance to tide and salinity are focal points that attract attention.

■ 海上森林－紅樹林。

攝影／薛美莉

沿著台灣的西海岸走一遭，您會在許多的河口地區看到一叢叢的海上森林。當人們第一次見到半淹在水中的森林時，總會想是怎樣的生命力可以在這樣的環境下生長，而當您知道那是片「紅樹林」時就更驚訝了，因為紅樹林竟是片綠油油的森林。在學術上的定義，紅樹林（mangrove）是指生長在熱帶及亞熱帶河口潮間帶的木本植物群落。「mangrove」是由西班牙文的樹（mangle）和英文的樹叢（grove）組合而成，取其生長型為密生叢林狀之意。中文名稱的由來則因為紅樹科植物體內含有大量的單寧（tannin），其木材常呈紅褐色，樹皮提煉出單寧為紅色染料，因此稱「紅樹」。

台灣有許多叫茄苳、茄萣地名的濱海城鎮或村落，這些地名有它的意義，那就是這裡有紅樹林。台灣是亞洲地區紅樹林分布上的邊緣地帶，對紅樹林植物而言，這裡的氣候有些冷，又因台灣沒有大河，它的棲地本極狹小，也因此在這塊土地上出現的紅樹林就更顯得珍貴。雖然台灣只有200多公頃的紅樹林，然而對於台灣整個沿海生態系的平衡、淨化河口污染物、防護海岸卻是極為重要，紅樹林的保育和復原就成為這個時代最重要的課題之一。

何處覓芳蹤

全世界紅樹林面積約有1,700萬公頃，主要分布在印度洋及西太平洋沿岸、美洲及非洲沿海，並以南北緯25゜度為主要分布範圍。世界分布的極限的最南為紐西蘭沿岸（44゜S），最北則為日本九州（31゜N），台灣位居太平洋西岸北緯22゜~25゜間，屬紅樹林自然分布之偏北地區。

紅樹林的生長除了要有熱帶氣候條件外，還必須符合具遮蔽性海岸線、廣大潮間帶及淺灘、維持正常進退的潮汐水流以及富含有機質的泥灘地等環境特性，才能讓紅樹林植物形成大面積群落。雖然台灣四面環海，但並非均適合紅樹林生長，台灣的海岸線可以簡單的分成東部斷層海岸、西部隆起海岸、北部沈降海岸及南部珊瑚礁海岸等4種地形。其中只有西部沿海具有潟湖、砂嘴及灘地等海積地形，才是紅樹林植物最理想的生育地。

根據早期文獻台灣紅樹林分布在基隆至屏東一帶河口，其中基隆灣和高雄灣的紅樹林在日治時期曾被指定為「天然紀念物」，高雄灣更是台灣最具

■淡水河口之水筆仔純林。

攝影／薛美莉

代表性的紅樹林地帶，除水筆仔外，其餘各種紅樹林植物均可發現，只可惜此兩處的紅樹林在基隆和高雄建港時，已不復存在。根據1992~1994年的調查資料，台灣紅樹林分布地點由南到北(台北淡水至屏東大鵬灣)，總面積約為287公頃。北部地區的淡水河口水筆仔純林是台灣面積最大的紅樹林區，共有北岸挖仔尾、南岸竹圍、關渡三個主要的生育地，目前均設立為「自然保留區」，受到良好保護。中部地區之紅樹林主要為防護灘地所栽植之海岸造林地，只有水筆仔及海茄苳兩種，但受強風影響多呈低矮灌木型。南部地區因熱帶氣候型再加上海灣潟湖地形，是台灣最適合紅樹林生長的地區。在南部沿海河口常可見胸徑40公分、高10公尺以上，估計樹齡在百年的海茄苳老樹，因此本區為目前台灣最具代表性的紅樹林生育地。

家族成員

根據國際紅樹林組織(International Society for Mangrove Ecosystem)所列，全世界約有24科30屬83種的紅樹林植物。若依其生育型可以分為真正紅樹林植物(true mangrove)和半紅樹林植物(minor mangrove)，真正紅樹林植物是指只生活在河口潮間帶之木本植物而且具有為適應環境而演化出之氣生根及胎生現象等，全世界約有60多種真正紅樹林。而半紅樹林則是指能在潮間帶生長亦能延伸到陸域生態系之植物。台灣真正紅樹林樹種計有4科7屬7種，分別如下：

- 紅樹科(Rhizophoraceae)
  - 紅茄苳(*Bruguiera gymnorrhiza* (L.) Lamk.)
  - 細蕊紅樹(*Ceriops tagal* C.B. Rob.)
  - 水筆仔(*Kandelia candel* (L.) Druce)
  - 五梨跤(*Rhizophora stylosa* Griff.)
- 海欖科(Avicenniaceae)
  - 海茄苳(*Avicennia marina* (Forsk.) Vierh.)
- 使君子科(Combretaceae)
  - 欖李(*Lumnitzera racemosa* Willd.)
- 大戟科(Euphobiaceae)
  - 土沉香(*Excoecaria agallocha* L.)

上述的7種植物中，紅茄苳及細蕊紅樹在台灣原本只分布在高雄灣一帶(現已開發為高雄港)，胡敬華1959年的調查報告中發現在高雄港區只剩22株紅茄苳和1株細蕊紅樹，此後就沒有此兩種植物的採集記錄，因此這兩種植物在近代海岸開發的過程中消失了。

■新竹紅毛河口為台灣海茄苳分布之北限。　攝影／薛美莉

適者生存

紅樹林生活於潮汐多變之河口、海灣、潟湖等鹽沼地，其生育地有土壤長期缺氧，鹽度變動極大等特性。一般植物無法於該環境下生存，在沒有競爭對手之情況下，使紅樹林得以成為優勢植物。然而，紅樹林植物為了適應嚴苛的環境挑戰，本身也發展出特殊生存方式來適應。

獨特的胎生現象

胎生現象是紅樹科植物顯著特徵，與一般植物的繁殖方式不同。當果實成熟時並不脫離母樹，而是由母樹上繼續供應養分萌發生長形成胎生苗，當胎生苗發育完全才脫離母樹。此種方法讓種子在發芽及小苗生長的過程中，避開惡劣的環境威脅。成熟的胎生苗組織富含間隙，比重較小，可以在海上漂浮，又因胚軸粗厚，富含單寧質，可避免海洋動物之侵襲，若不能順利固著亦可隨波逐流至較合適的生育地。此種傳播方式是紅樹林能廣佈熱帶海岸之主因。然而並非所有的紅樹林植物均有此特性，台灣的紅樹林植物中只有紅樹科的水筆仔及五梨跤有此胎生現象。

生命的固著點

在海岸河口的風浪侵襲及缺氧的環境下，使得紅樹林除了有土壤裡的根系外，也發展出一套適應環境的地上根。以台灣的紅樹林植物為例，水筆仔及五梨跤均會由上向下懸垂生出氣根，水筆仔多由莖基部發展出氣根並延伸為板狀支持根，而五梨跤則由分枝長出氣根且多於中途分歧，伸入土中，形成支持根，遠望猶如蜘蛛盤錯，所以在國外稱五梨跤為蜘蛛紅樹林(spider mangrove)。海茄苳之氣根由地下根縱走，向上長出散生的指狀呼吸根，其分布可達7~8公尺。欖李的屈膝根及土沉香的氣生根都是為了更方便吸收交換氣體。這些根系除能適應特殊環境，相對也增加了紅樹林的護岸功能，所以以往在漁塭岸經常栽植紅樹林作為護堤之用。

抗鹽英雄

紅樹林的生長環境土壤鹽分約在8~25‰間，在如此高鹽度的環境下紅樹林植物各自發展出一套拒鹽、排鹽的本領，以免植物體內累積過多的鹽分而造成毒害。大部分的紅樹科植物(水筆仔、五梨跤)其根系中含有高濃度的鹽分，所產生的負壓可以減少過多的鹽分進入體內，這類屬拒鹽植物。海茄苳葉片具有泌鹽腺，可分泌出植物體內多餘的鹽，所以下次看到海茄苳葉片上晶晶閃閃的結晶，不要懷疑，那就是鹽。

潮汐帶的生態舞臺

台灣的紅樹林面積約有287公頃，真正紅樹林樹種有4科5屬5種，若與全世界紅樹林相較在面積及種類上只佔少部分，然而長久以來，紅樹林在河口生態系中扮演著主要生產者的角色，經由枯枝落葉的分解，提供生物最基本的營養源及棲息孵育場所。如果少了紅樹林，那麼台灣西部沿海將成為毫無屏障的荒灘，而如果您有幸去拜訪紅樹林，那麼請您用雙眼仔細觀察，您會看到飛越樹稍的白鷺鷥，泥灘中辛勤覓食的螃蟹，緩行葉間的玉黍螺，彈跳的彈塗魚。或許您會覺得這樣的生態饗宴已經很豐富，但請注意這只是紅樹林沼澤中繁多生命的一部份，在這複雜交錯的生態系，您還有更多值得探索。

物種各論

紅樹林乃由不同科屬的植物所組成，其中以紅樹科為主要組成分子。又因其生長於海岸，種苗可以隨波飄流，世界性的分布極為廣泛。但紅樹林為熱帶海岸樹種，其生長受到溫度的限制。台灣位於紅樹林生長的邊緣地帶，冬季的低溫對紅樹林植物而言是一種嚴苛的考驗，因此組成樹種當然就不能和熱帶地區的相比，其中最佔優勢的水筆仔和海茄苳也是紅樹林家族中最耐寒的兩種。除上述真正紅樹林植物外，在台灣尚有些海岸植物如梧桐科（Sterculiaceae）銀葉樹（*Heritiera littoralis* Dryand.）、錦葵科（Malvaceae）黃槿（*Hibiscus tiliaceus* L.）、馬鞭草科（Verbcnaceae）苦林盤（*Clerodendrum inerme* (L.) Gaertn.）、玉蕊科（Lecythidaceae）穗花棋盤腳（*Barringtonia racemosa* (L.) Blume *ex* DC）、棋盤腳（*Barringtonia asiatica* (L.) Kurz）、千屈菜科（Lythraceae）水芫花（*Pemphis acidula* Forst.）、夾竹桃科（Apocynaceae）海檬果（*Cerbera manghas* Linn.）、豆科（Leguminosae）水黃皮（*Pongamia pinnata* (L.) Pierre *ex* Merr.）均列名於紅樹林組成分子。惟這些植物在台灣除苦林盤及黃槿外，其他種在台灣分布區域較狹隘，亦少在紅樹林區出現，下列則介紹幾種台灣的紅樹林植物。

■土沉香為半落葉性，其紅葉具觀賞價值。 攝影／薛美莉

眞正紅樹林植物

水筆仔 *Kandelia candel* Druce

紅樹科 Rhizophoraceae

常綠小喬木，樹皮灰褐色；葉革質對生，葉先端鈍，長5~10公分；聚繖花序，腋出成雙，每花基部有苞片1對；萼5裂，裂片線形；瓣5枚，每枚2裂之後，尚作絲狀細裂；雄蕊多數，花數細長；子房1室，柱頭3裂；果具宿存之萼。花期在6~7月間，胎生苗由12月起至翌年4月大量成熟，成熟胎生苗尖端呈紅褐色，胚軸長約15~20公分。水筆仔氣根多從樹幹下方發生，入地後即成支持根，自老樹幹基附近所發生之地下根，常有向地面隆昇而成板根者。

本種成熟時胚軸呈紅褐色，遠望似茄子，大陸稱為「秋茄」，主要分布於印度經馬來亞到我國東南部。台灣之分布由台北淡水河口、桃園、新竹、苗栗、台中、彰化、嘉義、台南、高雄沿海。在紅樹林植物中以水筆仔最能耐寒，因此在緯度較高的日本、福建及台灣北部均形成大面積純林。

■水筆仔之板根。

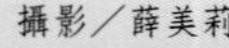

■水筆仔之胎生苗尤如綠色的筆。　攝影／薛美莉

五梨跤 *Rhizophora stylosa* Griff.

紅樹科 Rhizophoraceae

常綠喬木，樹皮灰褐而光滑；小枝粗大，落葉之後葉痕極為明顯；葉對生具長柄，橢圓形先端漸狹，長12~15公分；寬4~8公分；花具長梗，成聚繖花序，生於葉腋；萼及花冠均為4裂，花冠具有絲狀毛。裂片革質；雄蕊8枚，幾無花絲；藥為多室；子房2室。果實革質而具有宿存之萼。花期約在4~6月前後，胎生苗至翌年7~8月成熟。一般胚軸約15~25公分，成熟胚軸由綠轉深褐色且皮孔明顯。

五梨跤之主要特徵，為自樹幹及側枝之上端發生多數氣根。氣根於中途分歧，始入地下而成支柱根，常有因主幹枯萎而僅賴支柱根以支持其全體者。本種與水筆仔之分辨主要在其葉片較水筆仔大，尖端具芒狀突起，其胎生苗之胚軸較水筆仔長，且具皮孔。大陸稱本種為「紅海欖」，以往本種被誤為 *Rhizophora mucronata* Lam.已更正為 *Rhizophora stylosa* Griff.。

世界分布於非洲東岸、印度、馬來西亞、菲律賓群島至玻里尼西亞沿海。由於本種為嗜熱種，台灣只分布於嘉義、台南、高雄沿海。

■五梨跤開花。攝影／賴國祥

■五梨跤成熟之胎生苗。

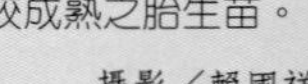

■五梨跤葉片較水筆仔大，尖端具芒狀突起。

■五梨跤之支持根具護岸功能。　攝影／賴國祥

海茄苳 *Avicennia marina* (Forsk.)Vierh.

海欖科 Avicenniaceae

常綠喬木，其高生長受生育地的鹽分及風浪影響極為顯著，樹形可為匍匐生長至大喬木，直徑可達40公分以上；樹皮白褐色，常作痂皮狀剝離；嫩枝有毛；葉革質，廣橢圓形，　面密布白色茸毛；花橘黃色，無梗，數朵簇生於枝之頂端；苞較萼短；萼作5裂，裂片凹面而具綠毛，花冠廣圓筒形，4裂，裂片平開；雄蕊4枚，具極短花絲而著生於花冠之喉部；子房4室，蒴果廣橢圓形，平扁，呈淡黃綠色而無毛；子葉2枚，疊摺而將轉軸包圍於其內；花期在5~7月間，果實約在9~10月成熟，具有隱胎生現象。

本種之主要特點為在徑幹之四周具有多數細長指狀呼吸根，係發自地下根；其呼吸根具海綿組織，對穩固生長及氣體交換極為有利。本種因樹皮為灰白色在大陸稱「白骨壤」。

分布印度、馬來西亞、澳洲、菲律賓、日本等地，是世界分布最廣的紅樹林植物。台灣分布北由新竹紅毛港南至屏東大鵬灣一帶，台灣南部的紅樹林以海茄苳為主要組成樹種，且樹形高大。

■ 海茄苳葉片上泌出之鹽結晶。

■ 海茄苳花為橘黃色，數朵簇生。

■ 海茄苳果實發芽。

■ 海茄苳的指狀呼吸根。

■ 海茄苳樹幹灰白色呈痂皮狀。

攝影／薛美莉

■欖李的果實細小，具厚纖維質。

■欖李開花時香氣濃郁為良好之蜜源植物。

其他紅樹林植物

欖李 *Lumnitzera racemosa* Willd.

使君子科 Combretaceae

常綠小喬木，高可達 25 公尺，樹皮呈褐色而粗糙；葉互生，肉質倒卵形，頂端圓形而略作凹陷，長約達3~4公分，寬1.5~2公分； 穗狀花序腋生，較葉為短，萼筒鐘形5裂，裂片呈三角型；花瓣 5 片白色；雄蕊 10 枚，成二列而著生於萼筒之上；子房1室；核果長橢圓形，長約1~4公分、寬約0.5公分，具厚纖維質有助果實飄浮水面。 本種一年有二次花期，一次為5~7 月，另一次於 10~11 月，主要以 5~7 月開花量較多。

分布熱帶非洲、印度、馬來西亞、菲律賓、澳洲、太平洋諸島、琉球、廣東。台灣原來分布於雲林、嘉義、台南、高雄沿海，以台南四草、南海寮及永安鹽田最大。

土沉香 *Excoecaria agallocha* L.

大戟科 Euphorbiaceae

常綠小喬木或具半落葉性，葉橢圓形，互生，長 5~10 公分。花為單性雌雄同株，雄花為葇荑花序，雌花有梗為短總狀花序。蒴果球形，有 3 深溝，熟時暗褐色，分裂3小乾果。本種因木材含樹脂燃燒具香味可為沉香代替品，故名為「土沉香」，其英名為 milky mangrove。大陸稱之為「海漆」概因其全株富含乳汁之故。

主要分布於印度、琉球、菲律賓、澳洲及廣東沿海。在台灣其主要分佈於西部沿海河流常與紅樹林混生。

■土沉香之地上根。

■土沉香之雄花。

■土沉香之具乳汁。

苦林盤 *Clerodendrum inerme* Gaertn

馬鞭草科 Verbenaceae

半蔓性灌木，嫩枝與花序均被有白毛；葉呈十字對生，厚革質，橢圓形，全緣；花序通常為3枚的聚繖花序，腋出或頂生；花萼杯狀，先端平截，具不明顯的5齒；花冠管狀，頂端5裂，白色，有時略帶紅暈；雄蕊4枚，著生於花冠喉部，花絲紫紅色，甚長，伸出花外；雄蕊1枚，子房上位，核果倒卵形。

苦林盤是沿海魚塭、河口及紅樹林區最常見的小灌木，其白色花冠及紫紅色的長花絲，很容易辨認，是一種世界性的紅樹林伴生植物。

■苦林盤為紅樹林最常見之伴生植物。

黃槿 *Hibiscus tiliaceus* L.

錦葵科 Malvaceae

多年生常綠喬木；樹皮灰色，嫩葉及花序有柔毛。葉厚紙質，心臟形，全緣或有小齒牙，背面灰白色，長 8~14 公分，寬 9~19 公分，掌狀脈 5~7 條；托葉早落。花頂生或腋生，具小苞；萼5裂，花瓣5，黃色，鐘狀，中心暗紫色；單體雄蕊，蒴果球形。

分布廣東、菲律賓群島、太平洋諸、印度和錫蘭等熱帶及亞熱帶海岸。台灣全島平野及濱海地區，是防風定砂的優良樹種。

■黃槿為半紅樹林植物。

攝影／薛美莉

※胎生現象

胎生現象為紅樹科顯著特徵，也是與一般植物最大的不同，水筆仔由開花至胎生苗成熟約需1年的時間。

● 下種

● 開花 5-6月

● 胎生苗成熟 3-5月

● 授粉 7-8月

● 下胚軸伸出 10月

● 結果 8-10月

攝影／薛美莉

海底來的訪客

◎撰文／曾喜育、伍淑惠、郭紀凡、邱文良
◎ Text ／ Hsi-Yu Tseng, Shu-Hui Wu
Ji-Fan Kuo, Wen-Liang Chiou

高位珊瑚礁

The Raised Coral Reef

高位珊瑚礁爲海底石灰岩沉積，經過造山運動及冰河期海平面升降作用所形成。台灣的高位珊瑚礁林主要分布在高雄柴山及恆春地區。尤其是位於恆春熱帶植物園後方的「墾丁高位珊瑚礁自然保留區」，該保

The raised coral reef is an accumulation of limestone from the sea bed formed through orogenic activity and rises from the sea level during the Ice Age. Taiwan's raised coral reef forests are mainly distributed on Kaohsiung's Chai Mountain and in the Hengchun area. A particularly noteworthy area is the raised coral reef forest behind the Hengchun Tropical Botanical

■盤結在高位珊瑚礁上的樹根。

留區內的林相以複層林爲主，並有豐富的藤本植物。土層堆積稍多處，植群的組成以黃心柿、大葉山欖、皮孫木、白榕、茄冬等較佔優勢，其中無論株數、底面積及林下的小苗數量，均以黃心柿爲最；在隆起的珊瑚礁岩塊上，則以紅柴、鐵色、白榕等爲主，白榕的植株常藉由支持根的支持，佔據廣大的生育地，形成「一樹成林」的景像。森林內的藤本植物也多彩多姿，目前觀察到的藤本植物種類有40餘種。保留區內植物繁茂，是許多野生動物生長活動的天堂。

Garden. The forest in this preserve is primarily mixed species, and there are abundant vines. The dominant species among the vegetation are *Diospyros maritima*, *Palaquium formosanum* Hayata, bird catcher tree (*Pisonia umbellifera*), weeping fig (*Ficus benjamina*), and *Bischoffia javanica* in the regions where the soil is accumulated more. Of these, in terms of the number of individuals, the ground coverage and the number of shoots below, sea hibiscus predominates; on the bulging blocks of coral, the Japanese cleyera (*Ternstroemia gymnanthera*), Philippine drypetes (*Drypetes littoralis*) and weeping fig predominate. The ficus trees often have extra, buttressing roots and take up the majority of the arable land, becoming a "one-tree forest". The vines in the forest come in all shapes and varieties. As of now, more than 40 species of vines have been observed. The lush vegetation in the preserve provides a heavenly habitat for many species of animals.

攝影／蕨類研究室

台灣森林的覆蓋面積約佔全島的58%，包涵了海岸、熱帶、亞熱帶、溫帶等各式各樣的森林形相。植物因其不同的生長習性、生育地（habitat）、傳播方式（dispersal）以及時間等錯綜複雜因子的交互影響，造就出不同的森林形相。位於台灣南部低海拔地區的高位珊瑚礁林（raised coral reef forest）為台灣森林中的獨特林相之一，由於特殊的地質環境以及地理位置的緣故，森林的組成種類與其它類林相有很大的差異。

台灣地區高位珊瑚礁範圍

高位珊瑚礁為海底石灰岩沉積，經過造山運動及更新世冰河期海平面升降作用所形成。台灣本島的高位珊瑚礁林主要有兩處，一處位於高雄的柴山地區，海拔達362公尺；另一處在恆春地區，其中以墾丁高位珊瑚礁林面積最大，保留也最完整。在地質環境相同、時間及氣候條件相近的情形下，兩處森林發育相近，植物種類組成也類似；但柴山因接近高雄市區，有人為遊憩及開發干擾的壓力存在，外來植物種類也較多。海拔範圍150~300公尺的墾丁高位珊瑚礁林，則在1994年受文化資產保存法保護，劃立為「墾丁高位珊瑚礁自然保留區」，面積有137.6公頃，目前交由林業試驗所恆春分所管轄，旨在保存完整的高位珊瑚礁林原始生態系和特有的石灰岩地質景觀。

高位珊瑚礁林探究

高位珊瑚礁原為海生動物珊瑚蟲之骨骼及貝類遺骸、海藻等沈積而成，地形處處艱險，高低起伏，且土壤層淺薄，究竟是如何形成原始森林呢？為了解這片奇特的生態環境，林試所及其他學術研究單位在此進行包括植物、動物、昆蟲及地質等等各方面的研究，希望能從蛛絲馬跡中解開此間錯綜複雜的關係。在植物方面，除以往陸續的植群組成分析研究外，林試所於民國83年開始設立永久樣區，以便了解森林的長期演替狀況。至90年12月已完成10公頃木本植物族群調查，其中記錄了100餘種，38,000多株的木本植物。

樣區內的林相以複層林為主，包括藤本植物及附生植物，植群之組成以黃心柿（*Diospyros maritima* Blume）、大葉山欖（*Palaquium formosanum* Hayata）、皮孫木（*Pisonia umbellifera* (Forst.) Seem.）、白榕（*Ficus benjamina* L.）、茄冬（*Bischofia javanica* Blume）等較佔優勢。無論株數或底面積，均以黃心柿為最高，且林下的小苗也佔了極大優勢；在隆起的珊瑚礁岩塊上，則以紅柴（*Aglaia formosana* Hayata）、鐵色（*Drypetes littoralis* (C. B. Rob.) Merr.）、白榕等為主，較高的礁塊則多為白榕的天下。這些植物未來的興衰與演替，值得深入研究。

■高位珊瑚礁上的植物。 攝影／蕨類研究室

特殊的藤本植物

森林內另一多彩多姿的成員則是藤本植物，雖然在研究上比其他樹種較少被觸及，但在植物生態系中，其生長模式卻是最有「行動力」的，只要一到達樹冠層，便不虞生長能源的匱乏。在樣區內雖沒有正式記錄藤本植物的種類，觀察到的已超過40種，以疏花魚藤（*Derris laxiflora* Benth）、亨利氏伊立基藤（*Erycibe henryi* Prain）、老荊藤（*Millettia reticulata* Benth）、腺果藤（*Pisonia aculeate* L.）等枝幹較粗。調查時最怕遇到的就是南天藤（*Caesalpinia crista* L.），它有另一個名字：搭肉刺，只要一被勾搭上，不免聽到一陣嘰哩咕嚕的咒罵聲，嚴重時尚可見血跡斑斑。但對植物來講，到底是人類闖入了它們的地盤，不過是小小的教訓一下罷了。

保留區的價值

保留區內植群繁茂，是許多野生動物生長活動的天堂，但也因而帶來殺機，過去常有獵捕者大量捕殺，以致從前曾在此活動的動物如山羌（*Muntiacus reevesi micrurus*）、穿山甲（*Manis pentadactyla pentadactyla*）等已不復見。且一有經濟價值，連植物也無法避免而受到牽連，原本尚有相當數量的象牙樹（*Diospyros ferrea* (Willd.) Bakhuizen），如今在林中也已寥寥無幾。常會聽到質疑保留區設立的意義，其實最大的好處是能讓這些地區得到喘息的空間，否則以目前台灣土地破壞的速率，福爾摩莎的美名將會很快的蕩然無存。本保留區緊鄰社頂村落，未來希望能朝向以社區力量來推動保育觀念，經由參與式管理方式，凝聚當地居民的力量與共識，保護這片美麗的環境。更願每個人都能自許為地球守護者，讓自然環境資源能延續給後代子孫。

■墾丁國家公園高位珊瑚礁植群。

攝影／蕨類研究室

物種介紹

■橢圓形的葉子，具有明顯的尾尖，使白榕呈現熱帶雨林物種的特性。

■一樹成林的白榕。

白榕 *Ficus benjamina* L.

桑科 Moraceae

屬於榕樹類植物，較大的枝幹常生出許多氣生根及衍生發育形成的支持根，是白榕特色。橢圓形的葉子，具有明顯的尾尖，使白榕呈現熱帶雨林物種的特性：當雨水過多時，雨水可由葉脈匯集，藉由導管似的尾尖將雨水宣洩，避免過多雨水停留在葉面上滋長微生物，造成葉子的腐敗。成熟紫黑色的榕果生於葉腋，常吸引鳥類的取食，經由其糞便將種子傳播至珊瑚礁岩上形成岩生植物；或掉落在其他種植物（宿主）上形成半附生性植物，最後因白榕強而有力的氣生根纏繞其宿主，造成宿主養分輸送困難而死亡，形成熱帶森林的殺手─「絞殺樹」。濃密的樹冠，加上由氣生根發育形成的支持根（有時具有絞殺性），使得樹冠下層難有其他大型木本植物之幼木可以存活；加上植株藉由支持根的支持，在本區域常佔具廣大的生育地，形成「一樹成林」的景像，可在恆春熱帶植物園白榕區欣賞此壯觀的景色。

■黃心柿的花和枝葉近照。

■黃心柿具有黃色漿果，大小約 2～3 公分，味道苦澀，野外少見動物取食。

黃心柿 *Diospyros maritima* Blume

柿樹科 Ebenaceae

高位珊瑚礁保留區內的森林組成樹種中，以小喬木的黃心柿最為優勢。因種子發芽率頗高，加上其幼苗具耐陰性，林下更新良好，使得黃心柿在保留區內擁有最大的族群。樹皮為黑褐色，皮孔多且粗糙，是生態調查時樹皮鑑定的特徵。漿果成熟時黃色，略球形，大小約2~3公分；味道苦澀，野外少見動物取食，所以種子多散佈於母樹週圍，因而常見有成群的幼苗在樹冠下。與毛柿（*Dispyros Philippensis* (Desr.) Gurke）、象牙樹（*Dispyros egbert-walkeri* Kosterm）同屬於柿樹科植物。萼片在開花後膨大不脫落（宿存），貼於漿果基部，是柿樹科與其他種類果實的最大差異。

茄苳 *Bischofia javanica* Blume

大戟科 Euphorbiaceae

本區的茄苳大多為巨木，幼齡木極少，這是先趨樹種（陽性）在較成熟林分的特色。因為先趨樹種的幼苗需要較多的陽光才能存活，這也是為什麼在茄苳樹冠下，常見到許多成千上百的種子發芽，但最後都死亡的原因。另外一方面，即使種子被鳥類或松鼠傳播在成熟林分中，但林下光度仍不足以

■茄苳開展的樹冠。

攝影／蕨類研究室

使其幼木生長，故在茂密的森林中只見茄苳的成熟林木。在恆春熱帶植物園中，常可見茄苳老樹盤踞在珊瑚礁岩上。樹形優美，栽植做庭園樹或行道樹非常適宜，著名的台東卑南綠色隧道即是茄苳！木材可供作傢俱、建築之用，但木材常腐朽而中空，使利用性大打折扣。新鮮葉子塞入雞腹中，可用來烹、烤美味的「茄苳雞」。花雌、雄異株，圓錐狀花序；雌花授粉後結果，一串串的像是褐色的葡萄。果實為漿果，成熟時吸引許多鳥類取食，亦可栽植作誘鳥植物。

■茄苳的果實。

■茄苳的基部。

大葉山欖 *Palaquium formosanum* Hayata

山欖科 Sapotaceae

常綠性的大喬木，具有豐富的白色乳汁，葉子倒卵形，革質，叢生於枝條頂端，是本種植物的特色。成熟植株具有熱帶樹種的特徵—板根，但不若銀葉樹的板根明顯。和人心果同為山欖科植物，核果狀的果實，橄欖綠，橢圓形，約4公分大小，於8、9月成熟，軟熟味甜可食，是老一輩童年時期甜蜜的野果，常吸引台灣獼猴、松鼠等取食。與茄苳等樹種在本區同屬於先趨樹種，種子發芽率頗高，常於母樹下見到小苗，但因陽光不足而少見發育至幼木。豐富的白色乳汁具有黏性，故又叫作「台灣膠木」。因樹形優美，且耐風，常被利用作行道樹及海岸防風樹種，台26線楓港至車城的路段所栽植的行道樹就有很多大葉山欖。

■大葉山欖的花苞。

■大葉山欖。

咬人狗 *Dendrocnide meyeniana* (Walp.) Chew

蕁麻科 Urticaceae

看名字就知道這植物不好惹！和咬人貓、蠍子草同為蕁麻科的植物。常綠的中、小喬木，葉子橢圓形，小枝、葉柄及葉兩面披有焮毛（刺毛）。葉子雖不若咬人貓、蠍子草那樣嚇人，但焮毛的毒性，若不小心碰觸則刺痛難耐，常達數小時至數天之久，不亞於前兩種植物中毒時的痛苦。不小心觸傷，可用肥皂水、濃茶或阿摩尼亞沖洗傷處；亦有人用姑婆芋擦敷，但因姑婆芋的汁液含豐富的植物鹼，對於過敏性皮膚反使該部位奇癢無比，形成二度傷害，應小心使用。圓錐狀的花序，在結實時，花托會膨大成白色肉質，晶瑩剔透，非常可愛，亦可以食用，味道微甜；若非具有惱人的焮毛，則會是一種非常不錯的觀果植物。木材的質地輕軟，恆春居民曾利用作魚網的浮標。

■咬人狗。

攝影／蕨類研究室

鐵色 *Drypetes littoralis* (C. B. Rob.) Merr.

大戟科 Euphorbiaceae

革質、彎彎的鐮刀形葉子，就是鐵色的特徵。本種分布台灣恆春半島及蘭嶼，分布狹窄、數量較少。果實成熟時紅色，三、五個著生於葉腋，非常好看，加上終年常綠，特殊的葉形，以及分枝多等特性，可以栽植成綠籬、庭園樹，或海邊防風樹種，是不可多得的原生觀賞樹木。唯本種生長緩慢，對於追求速成的台灣人而言，可能需要更多的耐性。至於名字的由來，可能是殷紅的果實遠看時像是鐵生鏽一般的紅褐色，故名叫作「鐵色」吧！

■鐵色。 攝影／蕨類研究室

皮孫木 *Pisonia umbellifera* (Forst.) Seem.

紫茉莉科 Nyctaginaceae

葉對生或輪生，長橢圓形，肉革質狀是皮孫木的特徵。「皮孫木」來自於其屬名「*Pisonia*」的讀音；小種名「*umbellifera*」則是形容皮孫木的繖形花序。台語「水冬瓜」暱稱則貼切的形容皮孫木的枝、葉像冬瓜一像，看起來肉肉、水水的。另外，皮孫木的英文名字叫馬來西亞捕鳥樹（Malay catchbird tree），非常有趣：可能因果實外具有腺體分泌黏液，鳥吃了果實黏住食道，使鳥無法再進食而死亡，或因翅膀被黏，無力起飛，易被人捉取而得名。植物的名稱不論是學名或俗名，常常具有其特色，把植物的特徵、習性或分布表露得一覽無遺；只要了解其中的義意，一般認為難認的植物，常可迎刃而解。另一種在保留區內常見的藤本植物—腺果藤則是本種的近親。

■皮孫木。 攝影／蕨類研究室

柿葉茶茱萸 *Gonocaryum calleryanum* (Baill.) Becc.

茶茱萸科 Icacinaceae

一個奇怪又拗口的植物名字，所謂「柿葉」意指其葉像柿樹的葉子。柿葉茶茱萸目前僅發現於恆春半島香蕉灣及墾丁高位珊瑚礁保留區內，為一局性分布之稀有種，廣泛分布於菲律賓呂宋島及巴丹島一帶，台灣為此種植物分布的北界。菲律賓當地居民採取其葉子，經過適當之處理，用以治療胃病。

其果實可飄浮於水面上，且無其他傳播者，推測本樹種可能是藉海漂自菲律賓傳播而來。隨著地質年代的變遷，當珊瑚礁隆起之際，柿葉茶茱萸由海邊分布到保留區內，香蕉灣海岸林為較晚時期隆起之地質，亦同時有此樹種之分布。本種雖在台灣分布侷限，族群數量也不多（約700株），但根據林業試驗所恆春分所近年來的研究，發現其族群遺傳之變異度相當高，結實豐富，林下有許多的小苗與幼樹，更新生長情形良好，目前並無滅絕之虞慮。

■柿葉茶茱萸。 攝影／伍淑惠

※榕屬（*Ficus* spp.）植物小百科

就植被地景來看，高位珊瑚礁森林無疑是恆春半島最特殊的景像；就植物的特色來看，台灣產榕屬植物近30種，除分布在蘭嶼、綠島的綠島榕（*F. pubinervis* Bl.）、安氏蔓榕（*F. trichocarpa* Bl. var. obtusa (Hassk.) Corner）、對葉榕（*F. cumingii* Miq. var. *terminalifolia* (Elm.) Sata）等，其餘種類皆可在位處熱帶的恆春半島發現，這可說是台灣榕屬植物的天堂。榕屬植物家族龐大，全世界約有近千種，多分布於熱帶及亞熱帶地區。具有白色乳汁、單葉多互生、葉芽為托葉包覆等雖是本屬的特徵；而亦花亦果的榕果，更是這個家族的最大特色。

榕果（fig）又稱作隱頭果（syncarp），其內部著生有多數小花，中空且膨大的花托（receptacle）(或花序軸)；人們因沒有看到榕樹開花就結果，便稱它為無花果。因為榕果幾乎密閉，只留小孔苞片形成的榕果小孔與外界相通，無法依靠蜜蜂、蝴蝶等授粉；加上雌、雄花成熟時期不同，無法自花授粉。然而，自白堊紀以來，有一群微小、特殊的授粉天使—榕果小蜂，隨著榕屬植物的演化，變成唯一可以鑽過榕果小孔為榕屬植物授粉的授粉蜂；歷經億萬年來的相互適應，彼此之間形成「種」對「種」的專一性授粉，進而達到高度的共生關係。

多樣化的生活型（life form）是榕屬植物的特色：榕樹（*F. microcarpa* L.）、白榕等具支持根的大樹（banyan），是高位珊瑚礁森林的常客；澀葉榕（*F. irisona* Elm.）、金氏榕（*F. ampelas* Burm. f.）及稜果榕（*F. septica* Burm. f.）等喬木常見於空闊、陽光充足、土壤較深厚的地區；相較之下，喬木型的大冇樹（*F. fistulosa* Reinw. ex Bl.）、幹花榕（*F. variegata* Bl. var. *garciae* (Elm.) Corner）則好生於溪谷潮濕的生育地；鵝鑾鼻蔓榕（*F. pedunculosa* Miq. var. *mearnsii* (Merr.) Corner）、斯氏榕（*F. tinctoria* Forst. f. ssp. *swinhoei* (King) Corner）等蔓性灌木，前者不懼高鹽的海水，後者不畏強勁的東北季風，悍衛著突起的珊瑚礁；台灣榕（*F. formosana* Maxim.）、菱葉濱榕（*F. tannoensis* Hay. var. *rhombifolia* Hay.）等小型灌木漸隱身在較內陸的森林中；大果榕（*F. aurantiacea* Griff. var. *parvifolia* Corner）、珍珠蓮（*F. sarmentosa* B. Hom.ex. J. Smith var. *nipponica* (Fr. et Sav.) Corner）等藤本植物，以莖上的不定根攀爬在其他喬木上攫取陽光，最具有「活動力」；具有半附生型的雀榕（*F. superba* (Miq.) Miq. var. *japonica* Miq.）、大葉雀榕(*F. caulocarpa* (Miq.) Miq.)等則在本區扮演林木殺手角色，只要被它們附生的植物，多半是難逃劫數，慘遭勒刑。

除了多樣的生活型，榕果著生位置從枝梢的腋生（leaf-axils）、老枝條的幹生（cauliflorous），以及經由特化繁殖的枝條深入表土的地下結果（geocarpic），也顯示了榕屬植物著果方式的歧異度(diversity)，有些種類同時擁有兩種以上的榕果著生方式呢。無論是哪一種著生方式，結實量大、軟甜多汁的果肉，對於鳥類、蝙蝠、靈長類等，引誘取食的吸引力非常大，榕果內微小的種子亦隨著這些傳播者到處傳播。在果實缺乏的季節，結果期長的果候（fruit phenology）特性，對於食果動物而言，是一項重要的食物來源，榕屬植物也因而有熱帶森林「關鍵種（keystone species)」之稱。

深入叢林

熱帶闊葉林

The Tropical Broadleaf Forest

◎撰文／楊國禎、謝長富
◎Text ／Kuoh-Cheng Yang
Chang-Fu Hsieh

台灣位處熱帶的邊緣，因受東亞季風的影響，冬季較爲寒冷，加上高山林立，及低海拔地區人爲的全面性開發，使得熱帶森林的面積大爲縮小，僅局限在南部及蘭嶼等地區。就物種組成及外貌而言，台灣的低海拔森林與典型的東南亞熱帶雨林有相當差異。由於台灣與南方熱帶地區並無陸地相連，最後冰河期結束後，由熱帶經台灣向北擴張的物種，以海飄及鳥類傳播爲主。且因北部氣溫較冷，由南而北熱帶物種的組成比例隨之減少。西南半部雖然氣溫較高，但因冬半季乾旱，耐旱與落葉樹種的比例較多。

Taiwan is situated on the fringes of the tropics, and as a result, it is impacted by the monsoons of East Asia. Winters are rather cold, and in the high mountain forests, and in the lower altitude areas that have been completely developed, the face of the tropical forest has been greatly diminished. Only isolated sections in southern Taiwan and on Lanyu island remain. In terms of plant makeup and appearance, Taiwan's low-altitude forests are quite similar to the typical Southeast Asian tropical rainforest. Given Taiwan's lack of a land connection with the southern tropics, at the conclusion of the last glacial periods, the plant species that passed through Taiwan on their way north from the tropics were mostly disseminated by wind or birds. Because the northern part of Taiwan is somewhat colder, the proportion of tropical species decreases as one moves northward. Although temperatures are somewhat higher in the southwest, the winter season is dry, and here there are more

■在恆春半島東部牡丹鄉的港仔溪上游，仍然保留著三百餘公頃的原始熱帶雨林，有一些大型黃藤突出樹冠。

蘭嶼島座落於東南外海，冬夏季風強勁，因此形相較佳的雨林均發育於避風的溪谷及山坡地。森林中來自呂宋島熱帶樹種的比例超過一半，在台灣地區高居第一。恆春半島東半部的溪谷及山麓地區也有林相優良的雨林，其中熱帶物種的比例較蘭嶼爲少。上述兩區森林的優勢樹種包括桑科、大戟科、茜草科等熱帶的類別，外觀呈現出熱帶雨林的部份特性，例如(1)幹生花現象，常見的有幹花榕、水同木；(2)兼具吸收養分與支持作用的支柱根，常見於正榕和白榕；(3)能增強根部固著作用的板根現象，以銀葉樹、幹花榕、大葉山欖和印度栲等較爲著名。其他森林中繁盛的天南星科植物、蕁麻科植物、蘭科植物、薑科植物與蕨類植物，包括有地生、著生與攀爬的種類，共同建構台灣熱帶雨林的景觀。

drought-resistant deciduous species. Lanyu Island is situated to the southeast of the island of Taiwan and experiences strong winds in the winter and summer. Consequently, the better examples of rain forests have sprung up in places that are protected from the wind, such as stream valleys or mountain slopes. In the forests, more than half the tree species come from Lusong Island, the highest proportion of such species in Taiwan. The stream valleys and mountain slopes of the eastern portion of the Hengchun Peninsula also have outstanding rain forests, but the proportion of tropical species is lower than in Lanyu. The preponderant species in the forests in the two areas described above include the Moraceae, the Euphorbiaceae, Rubiaceae, and other tropical varieties. On the surface this forest has some of the characteristics of a tropical rain forest, including (1) califlorous plants, the *Ficus variegata* Blume var. *garciae* (Elmer) Corner and *Ficus fistulosa* Reinw. ex Blume; (2) buttress roots for support and to absorb nutrients; as in the *Ficus microcarpa* L. f. and *Ficus benjamina* L.; (3) buttressed roots, which serve to increase a tree's stability, such as the *Heritiera littoralis* Dryand., *Ficus variegata* Blume var. *garciae* (Elmer) Corner, *Palaquium formosanum* Hayata, and *Castanopsis indica* (Roxb.) A. DC. The Araceae, Urticaceae, Orchidaceae, Zingiberaceae and fern grow luxuriantly in the forests, and include species that live on the ground and some which climb. Together, these plants make up the face of the Taiwanese rainforests.

攝影／楊國禎

■雀榕的葉芽。

攝影／桂曉芬

熱帶闊葉林的形相

北回歸線是北半球陽光年週期回歸直射的北界，也是地球表面所能吸取最多陽光能量的邊界。這條線恰好穿越台灣中部，致使地處低緯度的台灣低海拔地區終年溫暖無霜雪，均適合植物的生長。同時在雨量水分尚稱充足的情況下，很多樹木樹幹上的細胞都具有分化生長的能力，而形成常見的熱帶雨林景觀：

1. 花直接生長在樹幹上的幹生花現象，常見的有幹花榕（*Ficus variegata* Blume var. *garciae* (Elmer) Corner）、水同木（*Ficus fistulosa* Reinw. ex Blume）。
2. 代表空氣濕潤而生長的氣生根，當氣生根觸及地面後即變成兼具吸收養分與支持作用的支柱根，如此使一棵樹的樹幹很多、覆蓋面積很大，常見的有正榕（*Ficus microcarpa* L. f.）和白榕（*Ficus benjamina* L.）。
3. 樹幹基部橫向擴張及根部向上生長形成板根的現象，它能增強根部的固著作用，在地表淹水、土壤軟化的情況下，不至於倒伏，也不會因淹水而使根部缺氧死亡，如出名的銀葉樹（*Heritiera littoralis* Dryand），常見的尚有幹花榕、大葉山欖（*Palaquium formosanum* Hayata）和印度栲（*Castanopsis indica* (Roxb.) A. DC.）等。
4. 種子不落地也能發芽生長，因此在樹上、牆上和屋簷上形成附生或著生現象。隨後幼株的根系會迅速向下生長，伸入土中吸收養分而滋長，並逐漸的將著生母樹束住，當這棵母樹的樹幹逐漸加粗，主動的被纏勒，逐漸地當加粗到樹幹內的韌皮部被勒斷，養分無法由樹枝頂端送到根部，這棵母樹就會被勒死。常見的這類纏勒植物有雀榕（*Ficus superba* (Miq.) Miq. var. *japonica* Miq.）、大葉雀榕（*Ficus caulocarpa* (Miq.) Miq.）和正榕。
5. 氣候環境使植物生長迅速，一些種類的枝條常常枯死，在枯死枝條的基部因樹幹生長迅速而被樹皮包住，形成腫瘤，未被完全包住者形成樹洞，好像老態龍鍾的樣子，如茄苳（*Bischofia javanica* Blume）、正榕。
6. 葉子大都是大型的，也是潮濕大氣之下細胞能充分擴展的象徵。此類葉片通常具有尖尾巴，能迅速將葉面的水分傾洩掉，才不會因積水發霉而影響光合作用的能力。如澀葉榕（*Ficus irisana* Elmer）、大葉楠（*Machilus japonica* Sieb. & Zucc. var. *kusanoi* (Hayata) Liao）等樹種具有此特性。
7. 枯枝落葉會迅速分解，釋放出來的養分大都只存留於地表，植株如能在養分未流失前快速吸收，則對生長有利，因此雨林中優勢樹種的根部大都分布於地表，而成為所謂的淺根植物的現象。

■墾丁國家公園南仁山溪谷雨林中被灰莉（馬錢科）纏繞的大茄苳樹。

攝影／楊國禎

森林內的景觀

森林下繁盛的天南星科、蕁麻科、蘭科、薑科與蕨類等植物，構成溫暖與潮濕的象徵，包括有地生、著生與攀爬的種類：

1. 地生的種類—地上常密集生長、高約一公尺左右的大型草本植物，常見的有天南星科的姑婆芋（*Alocasia odora* (Roxb.) C. Koch）；蕨類種類最多，較顯著的有觀音座蓮（*Angiopteris lygodiifolia* Rosenst），雙蓋蕨類的廣葉鋸齒雙蓋蕨（*Diplazium dilatatum* Blume），樹蕨類的鬼桫欏（*Alsophila podophylla* Hook.）、台灣樹蕨（*Alsophila metteniana* Hance），實蕨類的尾葉實蕨（*Bolbitis heteroclita* (Presl) Ching）；蕁麻科植物的闊葉樓梯草（*Elatostema platyphylloides* Shih & Yang）及灌木的長梗紫麻（*Oreocnide pedunculata* (Shirai) Masamune）；薑科的有月桃（*Alpinia zerumbet* (Pers.) Burtt & Smith）、大輪月桃（*Alpinia uraiensis* Hayata）；蘭科的根節蘭類（*Calanthe* sp.），如寶島根節蘭（*Calanthe formosana* Rolfe）。
2. 著生的種類—著生樹幹上不同的部位，形成豐富而立體的多樣性內容，蘭科植物有大腳筒蘭（*Eria ovata* Lindl.）、豆蘭類（*Bulbophyllum* sp.）、羊耳蘭類（*Liparis* sp.）；蕁麻科的有烏來麻（*Procris laevigata* Blume）；蕨類的種類繁多，常見的為南洋巢蕨（*Asplenium australasicum* (J. Sm.) Hook.）、山蘇花（*Asplenium antiquum* Makino），大黑柄鐵角蕨（*Asplenium cuneatum* Lam.）、垂葉書帶蕨（*Vittaria zosterifolia* Willd.）、崖薑蕨（*Pseudodrynaria coronans* (Mett.) Ching）等。
3. 攀爬的種類—穿梭於林間，與著生植物共同組成攔截陽光、充分利用能源的角色，天南星科的有拎樹藤（*Epipremnum pinnatum* (L.) Engl.）和柚葉藤（*Pothos chinensis* (Raf.) Merr.）；蘭科的有台灣梵尼蘭（*Vanilla albida* Blume）。

■恆春半島雨林中的台灣野牡丹藤。 攝影／楊國禎

墾丁國家公園南仁山區熱帶雨林的代表性剖面圖

1.瓊楠	6.水同木	11.猴歡喜	16.單葉新月蕨	21.密毛小毛蕨
2.茄苳	7.石苓舅	12.姑婆芋	17.水藤	22.白鶴蘭
3.交力坪鐵色	8.山龍眼	13.冷清草	18.黃藤	23.中國穿鞘花
4.紅果椌木	9.雞屎樹	14.觀音座蓮	19.長穗馬藍	24.莎勒竹
5.白榕	10.九節木	15.翅柄三叉蕨	20.長花九頭獅草	25.台灣山蘇花

繪圖／范素瑋、廖啟政

乾濕與南北的差異

台灣位處熱帶的邊緣，又是東亞季風區涵蓋的範圍，冬季的東北季風為迎風坡的北部與東部地區帶來潮濕的季節，但相對較為寒冷，使之與熱帶稍疏離。然而被中央山脈與雪山山脈阻擋的西南部，處於背風面，是雨影區，因此冬半季相對乾旱。整體來說，台灣的低海拔森林與典型的熱帶雨林有所差別：東、北半部冬季稍冷，熱帶植物的組成比例降低；西南半部則冬半季乾旱，耐旱與落葉樹種的比例增多。兩區共同的特徵是小葉形樹種的比例較多。由於台灣與南方熱帶地區並無陸地相連，最後冰河期結束後，由熱帶向北擴張的物種，以靠海飄及鳥類傳播的為主。此類物種正隨著溫室效應的加強，由南端恆春半島登陸後向北擴展中。因而較典型的熱帶雨林景觀以南部及東部中央山區即將出山的河流兩岸、恆春半島的溪谷區以及地處東南外海的蘭嶼島較為顯著，但與東南亞以龍腦香科（Dipterocarpaceae）植物（台灣並無原生種）為主的典型熱帶雨林有相當差異。

■ 大腳筒蘭是南仁山雨林中最常見的附生蘭。

攝影／楊國禎

蘭嶼島座落於東南外海，不僅冬半季有強烈的東北季風，夏季也時常有強烈的西南氣流及颱風侵襲，因此形相較佳的雨林均發育於避風的溪谷及山坡地。蘭嶼島的森林中，來自呂宋島熱帶樹種的比例超過一半，在台灣地區高居第一。

恆春半島由於地理、氣候及歷史上的因素，乾旱的植群分布在西半部，從恆春到楓港、枋山之間。每到冬半季，籠罩在強烈的落山風之下。東半部自佳落水到旭海以北的地區，屬於迎風帶，雨量較多，因此避風的溪谷及山麓地區即發育出雨林來，孕育一些來自熱帶的物種，森林的優勢樹種包括桑科、大戟科、茜草科的類別。迎風的山頂部位則由樟科（Lauraceae）、殼斗科（Fagaceae）、冬青科（Aquifoliaceae）、八角茴香科（Illiciaceae）等東亞的代表性物種所佔據。

在蘭嶼的森林（2 公頃樣區）中以熱帶的樹種居大多數

成份	種數	百分比
熱帶	66	64.7
東亞（印度、中南半島、南中國）	14	13.7
日本及琉球	8	7.8
熱帶亞洲及東亞	1	1.0
蘭嶼特有	4	3.9
台灣及蘭嶼特有	9	8.8
合計	102	100.0

■ 熱帶
■ 東亞
■ 日本及琉球
■ 熱帶亞洲及東亞
■ 蘭嶼特有
■ 台灣及蘭嶼特有

在墾丁國家公園南仁山的溪谷森林（2.96 公頃樣區）中，熱帶的樹種遠較蘭嶼為低，東亞的物種及特有種的比例相對提高不少。

成份	種數	百分比
熱帶	32	29.4
東亞（印度、中南半島、南中國）	44	40.4
日本及琉球	5	4.6
熱帶亞洲及東亞	5	4.6
台灣特有	23	21.1
合計	109	100.0

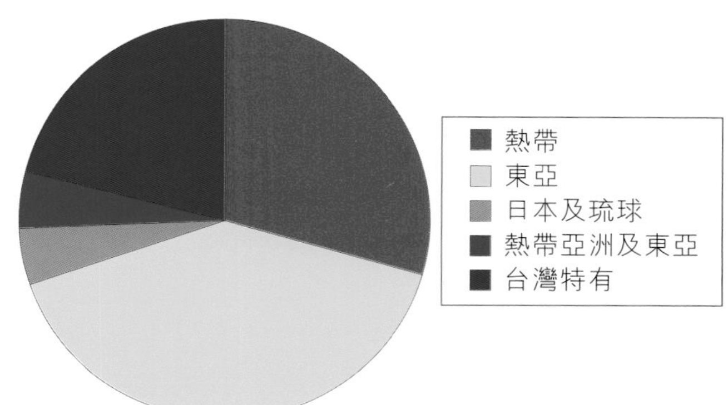

■構樹的果實。攝影／桂曉芬

消失的梅花鹿樂園

台灣西部丘陵地、西南部平原及花東縱谷，主要由山區崩瀉下來的砂石所堆積。砂石的保水能力差又貧瘠，加上經常性的氾濫沖積，在有人類活動的時期更因火耕、放火狩獵等等摧毀森林的情事，使得植被經常處於消長早期的草生地，或將演替為森林的過渡期中。尤其是西南部冬半年的相對乾旱期，使平原地區經常呈現出疏林草原的景觀。以苦苓（*Melia azedarach* L.）、莿桐（*Erythrina variegata* L.）、朴樹（*Celtis sinensis* Personn）、血桐（*Macaranga tanarius* (L.) Muell.-Arg.）、構樹（*Broussonetia papyrifera* (L.) L'Herit. ex Vent.）、山黃麻（*Trema orientalis* (L.) Blume）、白飯樹（*Flueggea virosa* (Roxb. ex Willd.) Voigt）等為主的獨立樹多散生在以甜根子草（*Saccharum spontaneum* L.）、芒草（*Miscanthus* sp.）為優勢的高草地上，成為草原性鹿科動物─梅花鹿的樂園。現今，這片土地已成為人類的住家、都市、工業區、田園、魚塭以及其他各型開發區，殘存的疏林景觀只散佈於堤防內的狹小的河床洪氾區、丘陵或平野的荒廢地，但梅花鹿已不復存在。

■著生在樹幹上的垂枝石松，其細長分枝的孢子囊穗向下懸垂，其上羽裂的附生蕨類為連珠蕨。 攝影／楊國禎

現今的植被概況

台灣島的人口密度將近世界第一，且主要集中於低海拔的平原及丘陵區，原先類似熱帶的植被區已被徹底更改過。通常土地利用類型由最自然到最人工化，可區分為原生林、次生林、人工林、果園（生產林）、草生地、田地、道路與建築。完全未遭受到破壞的原生林如前述，幾已不存。森林遭破壞後再長出來的次生林經常由血桐、構樹、山黃麻、白匏子（*Mallotus paniculatus* (Lam.) Muell. - Arg.）、稜果榕（*Ficus septica* Burm. f.）、蟲屎（*Melanolepis multiglandulosa* (Reinw.) Reich. f. & Zoll.）、小葉桑（*Morus australis* Poir.）、銀合歡（*Leucaena leucocephala* (Lam.) de Wit.）等樹種所組成，為處於過渡期的森林或灌叢。台灣南部與北部的次生林稍有不同，北部以山黃麻、白匏子為主，南部則以血桐、蟲屎、銀合歡等較多。由人工種植的樹林以相思樹、油桐類（*Aleurites* sp.）、桂竹（*Phyllostachys makinoi* Hayata）、刺竹（*Bambusa stenostachya* Hackel）等為主，其中桂竹多見於北部，刺竹則在中南部居多。果園（生產林）主要有柑桔類（*Citrus* sp.）、龍眼（*Euphoria longana* Lam.）、荔枝（*Litchi chinensis* Sonner.）、檳榔（*Areca catechu* L.）、芒果（*Mangifera indica* L.）、麻竹（*Dendrocalamus latiflorus* Munro）、綠竹（*Bambusa oldhamii* Munro）、椰子（*Cocos nucifera* L.）、釋迦（*Annona squamosa* L.）、柿子（*Diospyros kaki* Thunb.）等。草生地則以芒草（*Miscanthus floridulus*）、大黍（*Panicum maximum* Jacq.）、甜根子草、假儉草（*Eremochloa ophiuroides* (Munro) Hack.）、狗牙根（*Cynodon dactylon* (L.) Pers.）、盤谷拉草（*Digitaria decumbens* Stent）、巴拉草（*Brachiaria mutica* (Forsk.) Stapf）為主。田地則是人類的耕作區，隨著季節或年度而變更，水稻（*Oryza sativa* L.）、甘蔗（*Saccharum sinensis* Roxb.）與香蕉（*Musa sapientum* L.）曾經是過去最大宗的產物，玉米（*Zea mays* L.）、高粱（*Sorghum bicolor* (L.) Moench.）、甘藷（*Ipomoea batatas* (L.) Lam.）、花生（*Arachis hypogea* L.）以及各類瓜、果、蔬菜等傳統作物，一直是生活的必需品，但其生產在目前的多元化價值下，一直在變換中。

種類介紹

白榕 *Ficus benjamina* L.

桑科 Moraceae

大型喬木，高達 20 公尺，具有許多下垂的氣根，當氣根伸展觸及地面時即向下長根，而形成另一樹幹（支柱根），因此白榕多一樹成林，由許多粗細不等的樹幹支持著覆蓋面積廣大的樹冠層。白榕的樹皮呈灰白色，小枝下垂。葉片卵形至寬橢圓形，長 6~14 公分，兩面光滑，側脈由主脈向兩側平行展出。榕果常成對生長，球形或扁球形，直徑 1~1.5 公分，成熟時橘紅色。

白榕是亞洲熱帶的樹種，分布地包括馬來西亞、菲律賓、巴布亞新幾內亞、南太平洋諸島、印度、緬甸、泰囯、越南、大陸的雲南。並向北延伸至琉球群島及向南到達南半球的澳大利亞並北部。

在台灣，白榕是墾丁國家公園南仁山自然保護區內的溪谷中及蘭嶼雨林中的優勢樹種，在台東海岸山脈南段都蘭山一帶海拔500公尺以下的殘留森林中也能見到白榕。較易觀賞到的大白榕位在墾丁國家公園佳落水附近的林業試驗所港口工作站內。

■白榕的枝葉及榕果（隱頭花序）。

■墾丁國家公園港口地區的白榕支柱根。

■雀榕的枝葉及榕果。

■雀榕的榕果具有白色斑點。

■雀榕的葉芽。

雀榕 *Ficus superba* (Miq.) Miq. var. *japonica* Miq.

桑科 Moraceae

落葉性大喬木，具懸垂的氣生根，如同前述的白榕般的，會形成許多支柱根，也常著生於其他樹上，形成纏勒現象。每年落葉二至四次，新芽被大型的白色托葉包住。單葉互生，多簇生枝頂，呈橢圓形，兩面平滑，先端呈短尾狀，嫩葉呈紅褐色或茶褐色，葉柄長。隱花果球形或扁球形，有小梗，密生於小枝或粗幹上，初為綠色，後轉成淡紅褐色，外表有白斑點。雀榕的隱花果成熟時可以吸引成群的鳥類(如白頭翁、麻雀、五色鳥、綠繡眼、紅嘴黑鵯、珠頸斑鳩等）來啄食。

雀榕的原種分布在亞洲的熱帶地區，變種則分布於東亞地區。台灣全省的平地及山麓地區極為常見。

茄苳 *Bischofia javanica* Blume

大戟科 Euphorbiaceae

常綠或半常綠大喬木，高 20 餘公尺，胸徑最大的可達4公尺左右；樹幹圓直，一般主幹較短，分枝粗壯；樹皮灰褐色至棕褐色，樹皮粗糙，鱗片狀剝落；樹皮受傷後會流出紅色汁液。三出複葉，葉柄 8~16 公分長，小葉片卵形或卵狀橢圓形，長 6~12 公分，邊緣有小鋸齒。春天開花，花小，雌雄異株，形成圓錐花序。果實為漿果，圓球形，直徑 0.8~1.6 公分，淡褐色。

茄苳是亞洲熱帶及亞熱帶常綠雨林中的主要樹種，分布于印度、緬甸、泰國、柬埔寨、越南、馬來西亞、印尼、菲律賓、琉球群島、澳洲大利亞及和玻利尼西亞等處。

攝影／楊國禎

茄苳也是台灣700公尺以下森林中最常見的樹種，由於低海拔地區多已開發殆盡，僅少數巨木存在，約有66株的茄苳列入台灣鄉間珍稀老樹名錄，但其分布遍及全省。在殘留的低海拔熱帶森林中（如南仁山自然生態保護區及蘭嶼），茄苳及白榕均屬於最優勢的樹種。

■ 茄苳在春天發出的新葉及花序。

■ 長滿果實的茄苳樹。

■ 墾丁國家公園南仁山雨林中的茄苳樹。

大葉楠 *Machilus japonica* Sieb. & Zucc. var. *kusanoi* (Hayata) Liao

樟科 Lauraceae

常綠大喬木，葉片是台灣自生楠木類中最大的，長15~22公分，呈革質，長橢圓形或倒披針形，先端銳尖，上表面光滑有光澤，幼葉呈紅色。花序頂生，為聚繖圓錐花序，2~4月開花。果實扁圓球形，徑長1.2公分，夏末成熟，呈黑色。

大葉楠是台灣的特有種，全省低海拔原生林中的優勢樹種，喜生長於濕潤近溪谷地區。

■ 大葉楠的花序及新葉。

■ 大葉楠未成熟的果實。

■ 位於東部萬里橋溪畔的大葉楠森林，樹幹上有許多鳥巢蕨著生，並纏滿爬藤，林下則長滿蕨類植物及姑婆芋。

■ 樹頂佈滿新葉及花序的大葉楠。

攝影／楊國禎

■銀葉樹的花僅有花萼，無花瓣。
■銀葉樹的果實。

銀葉樹 *Heritiera littoralis* Dryand.

梧桐科 Sterculiaceae

常綠喬木，高可達20公尺，樹幹基部往往發育出明顯的板根。葉橢圓形至卵形，相當大，長15~36公分，幼葉紅色，成葉之葉下密佈銀白色鱗片及散生之褐色鱗毛。花單性，雌雄同株，無花瓣，聚成圓錐花序。果實扁橢圓形，長4~7公分，木質化，十分堅硬，背部有龍骨狀突起，內有氣室，能漂浮於海面散佈至各地。

銀葉樹是舊熱帶海岸林的組成樹種之一，東南亞、中南半島、菲律賓、琉球群島、西太平洋諸島、及東非均有其蹤跡。過去台灣北部的金包里（金山）、富貴角、基隆、宜蘭，南部的台南、恆春半島的牡丹灣、港口、龜子角、以及台東的成廣澳（三仙台北邊）、綠島及蘭嶼等地均盛產之，目前以墾丁公園海岸林及高位珊瑚礁上有較多的植株。

■銀葉樹具有巨大的板根。

攝影／楊國禎

咬人狗 *Dendrocnide meyeniana* (Walp.) Chew

蕁麻科 Urticaceae

咬人狗為常綠中型喬木，高5~7公尺，樹冠平截形，樹皮灰白色，光滑。幼枝、葉柄、葉背和花序都被覆短柔毛和刺毛，皮膚觸及時會痛癢發紅。咬人狗的葉子相當大，集生於枝頂，呈卵狀橢圓形，長達50公分，葉柄長10餘公分。花雌雄異株，呈圓錐狀聚繖花序。果實近圓形，由白色半透明的肉質果托所包圍。

咬人狗分布於全島低海拔平地及山麓地區，以東部及南部較多，屬於次生樹種。在熱帶林中，由倒樹或是人為干擾所形成的林隙中，往往被咬人狗所佔據。台灣地區本屬植物計有二種，在蘭嶼的森林中及海邊地區有另一種紅頭咬人狗，是蘭嶼的特有樹種。

■咬人狗具有大片的葉子。

構樹 *Broussonetia papyrifera* (L.) L'Hérit ex Vent.

桑科 Moraceae

又稱之為鹿仔樹，是落葉中喬木；全株密生短毛。葉具長柄，呈心狀卵形，有時3或5深裂，長10~20公分，表面組糙。花單性，雄花聚成下垂的柔荑花序；雌花則密生成圓球形，有毛。聚合果球形，徑約2公分，熟時缸色，可食用。

構樹廣泛分布於馬來西亞、中南半島、印度、中國大陸南方、太平洋諸島等地。在台灣自平地以至山麓地帶的次生林內或砍伐地，均常見之。本樹種之樹皮多纖維，在過去常用以造紙，樹葉也是家畜的飼料。荷蘭時代以前，廣大的嘉南平原主要由芒草地夾雜次生林的植被類型所覆蓋。快速生長的構樹即是梅花鹿大量的食物來源。

■構樹的雄花序。

■構樹的圓球形雌花序及成熟的果實。

血桐 *Macaranga tanarius* (L.) Muell.-Arg.

大戟科 Euphorbiaceae

常綠中喬木，高5~8公尺，枝幹斷裂後會流出汁液，乾　變紅，故稱為血桐。嫩葉及幼枝被有柔毛，小枝粗壯。葉片近圓形或卵圓形，長15~30公分，基部鈍圓，盾狀著生，葉上面無毛，下面密生白毛，掌狀脈9~11條。花單性，雌雄異株，無花瓣，長於花苞中；雄花序圓錐狀，雌花較少，叢生。蒴果由2~3個分果合成，密被顆粒狀腺體及軟刺。

血桐廣泛分布於低海拔平野及低山地區，屬於不耐遮陰的先驅樹種，生長快速。目前台灣的平原地區，不論在河床及河堤上，或是久經荒廢的耕作地，最先長出的樹種，多半是血桐或前述的構樹。

血桐是熱帶廣佈種，分布於馬來西亞、印尼、菲律賓、澳洲北部、越南、泰園、緬甸及琉球群島等地。

■ 血桐的雄花序及盾狀葉片。

相思樹 *Acacia confusa* Merr.

豆科 Fabaceae

常綠大喬木，高可至15公尺，全株無毛。幼苗時期長出的第一片葉子為一回或二回羽狀複葉，這是真正的葉子，隨後發出的葉片是由葉柄扁平化所形成假葉，呈革質鐮刀狀或披針狀，長6~12公分。花金黃色，春夏或秋冬間開放，有微香，集合成形成圓球形的花序。莢果扁平，長5~10公分。

相思樹原產於菲律賓、台灣及蘭嶼，恆春半島原生。在過去瓦斯尚未普遍的時代，是台灣最重要的薪炭材來源，曾大量栽植，為海拔1,000公尺以下地區最常見的樹種。相思樹生長迅速，加上本身屬豆科植物，根部具固氮能力，能在最貧瘠的紅土台地自行繁殖，因此也成為桃園、中壢、大度山、八卦山等台地的最佳防風及遮蔭樹種。

火筒樹 *Leea guineensis* G. Don

火筒樹科 Leeaceae

灌木或小喬木，高3~6公尺。葉為三至四回羽狀複葉，全長50~80公分，大小葉柄基部均有關節，小葉對生，卵形至橢圓形。火紅的大型花序在雨林萬綠間極為顯眼，但每朵花極小，花瓣呈黃白色。果實為扁球形的漿果，剛開始結果時呈淺綠色，繼而轉為土黃色、暗紅色，最後成熟時為黑紫色。

火筒樹這類植物產於非洲、馬達加斯加、泰國、中國雲南、菲律賓等處，台灣多見於恆春半島東半部、蘭嶼的雨林中，向北分布到高雄六龜等處的低山地帶。

■ 蘭嶼雨林中的火筒樹。

■ 相思樹有著金黃色的球狀花序。

■ 結滿豆莢的相思樹。

攝影／楊國禎

■ 柚葉藤成熟的紅色漿果。

柚葉藤 *Pothos chinensis* (Raf.) Merr.

天南星科 Araceae

蔓性多年生常綠藤本，綠色的莖細長多節，常呈之字形彎曲，葉柄有翅（單身複葉），與柚子葉類似，葉片本身呈卵狀披針形，長5~10公分。肉穗花序腋生，基部由4~5片綠色的苞片所包覆。橢圓形的漿果長約1公分，成熟時為紅色。

柚葉藤分布於中國華南及琉球。在台灣低中海拔地區均可見到，特別是低海拔的闊葉林內極為普遍，著生在樹幹及岩石上。

■ 柚葉藤的圓球形肉穗花序，基部有數片綠色的苞片。

大果榕 *Ficus aurantiaca* Griff. var. *parvifolia* (Corner) Corner

桑科 Moraceae

大型常綠攀緣性木質藤本；葉厚革質，倒卵形至橢圓形，成熟粗壯的結果枝上的葉子長3~6公分，葉下表面具格子狀斑點，並有毛絨。隱花果相當大，呈球形或橢圓形，長5~7公分，外表略有絨毛，成熟呈橘紅色，可食用。

大果榕是東南亞熱帶雨林常見的物種，多隨樹幹攀爬至四、五十公尺高的樹冠之上。在台灣僅分布至恆春半島、蘭嶼及綠島的雨林之中。

■ 大果榕的榕果呈橘紅色。

■ 攀附在枯樹枝幹上的大果榕。

黃藤 *Calamus quiquesetinervius* Burret

棕櫚科 Arecaceae

多年生大型藤本，莖長達數十公尺，粗5~6公分，密生長刺。葉為羽狀複葉，3.5公尺長，葉柄及葉脈上均有倒刺，有時葉柄先端會特別延伸呈鞭狀，滿佈倒刺但不長羽葉。花雌雄異株，聚成圓錐花序。果實橢圓形，長約2公分，外表覆有14~17排鱗片，成熟時呈黃色。

黃藤為台灣及蘭嶼的特有種類，在低海拔闊葉林內極為普遍。過去曾大量採伐，抽取藤心，以製作傢俱。目前各地森林中之族群數量已逐漸恢復。

■ 攀附至樹冠層的大黃藤。

■ 黃藤的果實被覆白色的鱗片。

攝影／楊國禎

南洋山蘇花

Asplenium australasicum (J. Sm.) Hook.

山蘇花類又稱為鳥巢蕨類，多附生在樹幹及岩石上，喜好潮濕的森林。葉片帶狀叢生，致使植株呈鳥巢狀；帶狀葉片長1~1.5公尺，側脈平行，孢子囊群沿葉背面的側脈生長。與其他山蘇花類的台灣山蘇花（*Asplenium nidus* L.）、山蘇花（*Asplenium antiquum* Makino）最容易的區別，在於葉背中肋有龍骨狀突起。

南洋山蘇花廣泛分布於亞洲熱帶地區。在台灣分布於低海拔之山區及綠島、蘭嶼。

山蘇花類植物的葉形優美，是插花時最好的陪襯材料。其嫩葉先端呈蜷曲狀，質地嫩脆，無苦澀味道，適合作為蔬菜食用，是目前最流行的野菜，花蓮地區首先大規模栽培。

著生在樹幹上的台灣山蘇花，具有寬大帶狀的葉片，長出淡紅色下垂花序的植物為台灣野牡丹藤。

南洋山蘇花。

攝影／蕨類研究室

姑婆芋 *Alocasia odora* (Roxb.) C. Koch

天南星科 Araceae

多年生大型草本，高可達2公尺以上，地下莖粗大。葉盾形，葉片寬卵形，長60~100公分，基部心形；葉柄甚長，達到1.2公尺。佛焰花序春夏間發出，佛焰苞下半段綠色，上半段長橢圓狀披針形，黃白或黃綠色，長15~25公分；佛焰苞中間之穗軸肉質，雌花在下，雄花在上，中央為無性花。漿果球形，多數，聚生在果軸上，成熟紅色。

姑婆芋廣泛分布在東南亞及東亞地區，從印尼北至日本的九州均極普遍。生長在熱帶、亞熱帶雨林，乃至暖溫帶常綠闊葉林下，每成優勢種類。因全株有毒，根、莖、葉皆不可食。

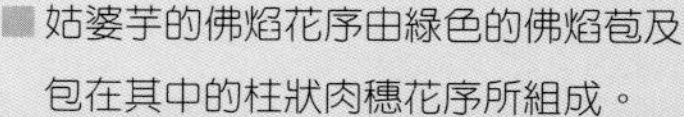

姑婆芋的佛焰花序由綠色的佛焰苞及包在其中的柱狀肉穗花序所組成。

姑婆芋成熟的果實為紅色的漿果。

雨林下常見的姑婆芋，具有寬大的葉片，前方葉脈平形的植物為仙茅科的船仔草。

攝影／楊國禎

亞熱帶的容顏

樟楠林

The Low-Altitude Broadleaf Forests

◎撰文／陳子英
◎ Text ／ Tze-Ying Chen

低海拔的闊葉林主要分布於台灣中部海拔到500~1,500公尺，在北部則分布於50~900公尺，組成上以樟科的木薑子屬和楠木屬、桑科榕屬及殼斗科為主；景觀上多為常綠闊葉林，但在地形上仍有區分，在山坡的多為殼斗科優勢的櫧木型，而在溪谷主要以大葉楠為主的楠木型，由於全台灣地區季節雨量的不平均、過去植物的遷移歷史和地質的影響，使得山坡型的櫧木類在台灣地區的植物有明顯的分化，這些優勢種的不同，使得台灣各區植物社會有所區別，並豐富了台灣低海拔地區不同區域的植物多樣性。台灣地區雨量充沛在這個林帶的窪地及不透水層地層常出現湖沼濕地，同時在東北角附近稜線上東北季風強盛因此局部地區有風衝矮林；低海拔地區由於不斷的被開墾及伐木，再加上河川常向上游侵蝕形成小面積的崩蹋，次生植群及早期的芒草成群生長，早期薪炭材及桐油的需求，許多地區栽植相思樹、油桐、果樹、茶園及農作物，或開發山坡地成為城鎮。台灣地區低海拔使用率較其他林帶高，殘存的原生森林不多，我們應好好保存這片森林。

■東北角的草嶺古道附近，冬季東北季風之影響，植群避至山凹處。 攝影／陳子英

The low-altitude broadleaf forests are primarily distributed throughout central Taiwan at altitudes of 500 to 1,500 meters. In the northern part of the island, they occur at altitudes of 50 to 900 meters. Plants here are mainly Lauraceae family plants such as *Litsea* and *Machilus* genus, Moraceae family ficus species and Fagaceae. The view here is primarily one of broadleaf evergreens, but there are still areas in the topography where the *Castanopsis* type are the predominate species on the hillsides, and in the stream valleys, the main species is *Machilus japonica* Kusanoi. Since different places in Taiwan have differences in seasonal rainfall and different histories of immigrant plants in the past, sweet oak is evident on the mountain slopes of Taiwan. The differences in the dominant species cause differences in the plant communities in different areas in Taiwan, and enriches the diversity of plant life in different areas of Taiwan's low-altitude regions. The rainfall in Taiwan accumulates in the hollows on this forest belt and on impermeable layers, and often produces marshes. At the same time, the northeastern corner sees shorter plants due to the impact of the winds, and because of the higher frequency of land use at the lower elevations, plus natural interference, in these areas we can see secondary vegetation or forest and early silver grass (*Miscanthus* species) growing. Because of the earlier need for fuel and tung oil, many areas were planted in Formosan koa (*Acacia confusa*), tung oil trees, fruit trees or crops, or cleared to provide areas for villages and towns. There is more usage of Taiwan's low-altitude areas than other areas of the island, and there are not many primitive forests remaining. Those that are left must be conserved well.

生育環境

低海拔樟楠林在海拔分布上中部地區約為500~1,500公尺之間，北部地區則降到50~900公尺，雖然雨量在全台灣分布不均勻，但整體而言，年平均降水量尚稱充足，冬季並無降霜的現象，樹幹上較少有像中海拔闊葉林樹幹上掛滿生長繁盛的苔蘚類植物。由於季節雨量的不平均使得低海拔樟楠林的生育環境出現明顯的差異，在北部地區冬季潮濕多雨，西南部呈現乾旱，東部地區則呈現二種氣候並存的現象，在受到山脈阻擋的地區呈現乾旱，沒有山脈阻擋的地區則明顯乾燥；由於氣候差異所形成的分隔使得各區的森林中各有不

■低海拔闊葉林森林一景。

攝影／陳子英

■蘇澳附近之筆筒樹林。 攝影／陳子英

同的優勢樹種存在。樹種的組成上除南部地區及東部地區較乾旱之地區有落葉的闊葉林外，大多屬於常綠闊葉林為主，由於低海拔地區開發的時間較長，且屬於人類活動頻繁的地區，這些殘存的原生或次生植物社會多分布在陡峭之坡地或山坡稜線上，甚至部份保留在軍事基地之中。

在結構上的分層，可分成幾個層次，上層的優勢樹木由許多樟科的樟屬（*Cinnamomum*）、楠木（*Machilus*）、桑科榕屬（*Ficus*）或殼斗科的植物所構成，樹冠層的高度在溪谷較高，通常會達到15~25公尺，在中坡或稜線較低矮，通常只有5~10公尺；在中下層則有烏皮茶（*Tutcheria shinkoensis* (Hay.)Nakai）、山龍眼（*Helicia formosana* Hemsl.）、樹杞（*Ardisia sieboldii* Miquel）、茶科之柃木屬（*Eurya*）、桑科榕屬植物、雞屎樹（*Lasianthus*）、柏拉木（*Blastus cochinchinensis* Lour.）、紫金牛科（Myrsinaceae）植物等組成，部份桑科榕屬植物的隱頭花序會直接生長在樹幹上；在上層優勢樹木的樹幹上常常有許多植株大小不同的附生植物，如南洋巢蕨(*Neottopteris australasica* J. Sm.）、台灣山蘇花（*Neottopteris nidus* (L.) J. Sm.）及伏石蕨（*Lemmaphyllum microphyllum* Presl）等蕨類及蘭科（Orchidaceae）植物，而下層草本層主要有蕨類的雙蓋蕨（*Diplazium*）及觀音座蓮（*Angiopteris lygodiifolia* Rosenst.）、薑科（Zingiberaceae）、蘭科與禾本科植物等，其中著名的報歲蘭（*Cybidium sinensis* (Jackson *ex* Andr.) Willd.）就生長在本區。

生育地的分隔

低海拔的原生樟楠林依地形位置可分成二種不同的植物社會，一種是分布在山坡或稜線上以殼斗科植物為主的山地型植物社會，又稱為櫧木型植群（*Castanopsis* type），櫧木型植物社會主要優勢種為殼斗科植物，但依各地乾旱季節的差異，出現不同的殼斗科優勢植物，所謂殼斗科植物是一群果實外面由殼斗包被的木本植物所構成，這一類的植物果實通常為松鼠等嚙齒類動物最喜歡的食物；另一個植物社會則是以樟科的楠木類及桑科榕屬為主的溪谷植物社會，稱為楠木型植群（*Machilus* type），楠木型植物社會主要以大葉楠（*Machilus japonica* Sieb. & Zucc. var. *kusanoi* (Hayata) Liao）為主，桑科榕屬之植物則有九重吹（*Ficus nervosa* Heyne *ex* Roth.）、澀葉榕（*Ficus irisana* Elm.）、菲律賓榕（*Ficus ampelas* Burm.f.）、豬母乳（*Ficus fistulosa* Reinw. *ex* Bl. f. *fistulosa*）及大有榕（*Ficus septica* Burm.f.）等，並伴生有九芎（*Lagerstroemia subcostata* Koehne）及茄苳（*Bischofia javanica* Blume）等溪濱植物。

若依氣象的播報圖將臺台灣地區分成東北區、西北區、中西區、西南區、東南區及東部地區，由於各區氣候乾旱季節的差異、過去植物的分布與遷移及地質的影響，使得台灣低海拔闊葉林各區的植物社會組成有些微的差異，這些差異可以由山坡這一型植物社會樹冠層優勢種之殼斗植物的分布看出端倪。若將只出現在某些區的殼斗科植物放入各區的框架中，如東北部之低海拔地區的風衝矮林，主要優勢種有白背櫟（*Cyclobalanopsis salicina* (Bl.) Oerst），東部地區有石灰岩之地質的太魯閣櫟（*Quercus tarokoensis* Hayata）及除東部地區外也分布於東南區的星刺栲（*Castanopsis fabri* Hance）與灰背櫟（*Cyclobalanopsis hypophaea* (Hayata) Kudo）；西南區及東南區有台灣苦櫧（*Castanopsis formosana*(Skan)Hayata）、印度苦櫧（*Castanopsis indica* (Roxb.) A. DC.）、細刺苦櫧（草野氏櫧）（*Castanopsis kusanoi* (Hayata)）、嶺南青剛櫟（*Cyclobalanopsis championii* (Benth.) Oerst）、；中西區有火燒柯（*Castanopsis fargesii* Fr.）及反刺苦櫧（*Castanopsis eyrei* (Champ. *ex* Benth) Hutch.）；東南區有柳葉石櫟（*Pasania dodonaeifolia* Hayata）、浸水營石櫟（*Pasania shinsuiensis*

■單刺苦櫧的白色葇荑花序。 攝影／陳子英

棲蘭附近低海拔闊葉林。 攝影／陳子英

(Hayata & Kanehira) Nakai)、加拉段石櫟(*Pasania chiaratuangensis* (Liao) Liao)及台灣石櫟(*Pasania formosana* (Skan *ex* Forbes & Hemsl.) Schottky),部份的殼斗科植物或出現在少數幾個地區中,如單刺苦櫧(白校欑)(*Castanopsis cuspidata* (Thunb.*ex* Murray) Schottky var. *carlesii* (Hemsl.) Yamazaki f. *sessilis* (Nakai) Liao)及烏來柯(*Castanopsis uraiana* (Hayata) Kaneh. et Hatus.)只出現東北區及中西區,油葉石櫟(小西氏石櫟)(*Pasania konishii* (Hayata) Schottky)分布於中西區及東部地區,菱果石櫟(*Pasania rhombocarpa* (Hanace) Schottky)及後大埔石櫟(*Pasania cornea* (Lour.) Liao)出現於中西區、西南區及東南區,捲斗櫟(*Cyclobalanopsis pachyloma* (O.Seem.) Schottky)只出現在東北區、中西區及東南區,南投石櫟(*Pasania nantoensis* (Hayata) Schottky)只出現中西區及東南區。

由於這些殼斗科植物都是森林中的主要優勢種,林下的伴生植物及其他生物與它們也有密切的相互關係,各區域的差異,就如同在台灣低海拔的樟楠林這一層公寓內做了許多隔間,隔間中的植物種類略有不同,因此豐富了低海拔物種的歧異度。

次生植群及森林

低海拔闊葉林由於靠近人類活動的地區,因此不斷的被開墾、伐木及焚燒,加上亞熱帶地區雨量充沛,臺灣地區山勢陡峭,河川常向上游侵蝕,因此偶有小崩塌地產生,這些被人為干擾的山坡或崩塌地在初期常形成以芒草為主的草地,漸漸由演替初期的先鋒植物如白匏子(*Mallotus paniculatus* (Lam.) Muell.-Arg.)、山黃麻(*Trema orientalis* (L.) Bl.)、血桐(*Macarange tanarius* (L.) Muell.-Arg.)、野桐(*Mallotus japonicu* (Thumb.) Muell.-Arg.)、粗糠柴(*Mallotus phillippinensis* (Lam.) Muell.-Arg.)、蟲屎(*Melanolepis moluccanun* (Reinw.) Reich.f.& Zoll)、小桑樹(小葉桑)(*Morus australis* Poir.)及杜虹花(*Callicarpa formosana* Rolfe)等植物漸漸進駐,這些植物生長迅速,漸漸取代芒草草地,在北部潮濕且向陽的地區並夾雜有筆筒樹(*Cyathea lepifera* (J. Sm. ex Hook.) Copel)的群落,甚至在河川的開闊地有台灣赤楊(*Alnus formosana* (Burk.*ex* Forbes & Hemsl.) Makino)分布,例如在開闊的蘭陽平原上,早期就存在許多的台灣赤楊純林,地名上稱為柯仔林(台灣赤楊俗名柯仔)。

經過幾十年的演替後,這些陽性的先鋒植物由於種子無法在自己森林下生長,因此漸漸被陰性的樟科植物、桑科榕屬植物或殼斗科的櫧木類所取代,隨後又回復成以樟科的楠木類、桑科的榕屬植物或殼斗科為主的低海拔闊葉林。然而由於低海拔地區靠近人類居住的地區,常常有人為及自然的干擾,因此次生植群及森林就隨處可見。

■南澳南溪附近低海拔闊葉林。 攝影/陳子英

人造相思樹林及油桐

低海拔闊葉林在日治時代與台灣光復初期一方面為解決民生燃料的問題,另一方面在低山之丘陵地或陡峭山坡地為了水土保持的考量,廣泛栽植分布在恆春半島及菲律賓北部的相思樹做為薪炭材及水土保持林,然而近年來由於環境保護及追求低污染的雙重考量下,以天然氣取代薪炭材成為民生重要的能量來源,低海拔地區不再建造相思樹林,現存大多數的相思樹人工林亦不再進行撫育,旁邊天然或次生林的闊葉樹小苗又漸漸的進入相思樹人工林中,使得許多相思樹人工林呈現天然闊葉樹及相思樹混雜的景觀,混雜的闊葉樹主要有樟科的香楠(*Machilus zuhoensis* Hayata var. *zuihoensis*)、豬腳楠(*Machilus thunbergii* Sieb & Zucc.)、黃肉樹(*Litsea hypophaea* Hayata)等及桑科榕屬的豬母乳、大冇榕及小葉桑,與陽性的植物如山黃麻、杜英、江某、白匏子及野桐等。

相似於相思樹,早期因桐油及木屐需求而引進中國大陸的油桐類植物,如千年桐(*Aleurites montana* (Lour.) Wils.)及三年桐(*Aleurites fordii* Hemsl.),建造成油桐林,但50年代後由於桐油已可使用化學合成,加上目前的木屐需求量較少,使得油桐類人工林已變成與闊葉樹混雜的景觀,但由於油桐的白花開花成簇,在中部或南部地區落葉前變黃色,因此每逢春天開花,或冬天落葉前都成為低海拔淺山的特殊景緻。因此在台北縣某些地區油桐開花時,成了當地的重要節慶稱為「油桐季」。

種類介紹

桑科榕屬植物

島榕*Ficus virgata* Reinw. *ex* Bl. var. *philippinensis* (Miq.) Corn.

桑科 Moraceae

桑科榕屬的植物，花與無花果相似，屬於隱頭花序；所謂隱頭花序，是指花序由外面的苞片整個將花朵包住，同時花授粉形成果實後，果實也不開裂，因此花與果實較難區分，但島榕果實成熟後顏色轉變成紅色，因此花和果實的區別較容易；它的更新方式比較特殊，通常是小苗附在半遮蔭的岩石或粗糙的筆筒樹樹幹上，再慢慢將幾條氣生根長到地面上，當它的氣生根接觸地面上時就會形成主根，等這些根莖長大後會將附生的植物纏勒至死，因此也是一種纏勒植物。

攝影／陳子英

豬母乳 *Ficus fistulosa Reinw. ex* Bl. f. *fistulosa*

桑科 Moraceae

屬於常綠喬木，花為隱頭花序，並常在樹幹上開花，有樹幹長花的現象，花外為綠色苞片所包住，當果實成熟時顏色略轉成褐色，與大冇榕都是低海拔闊葉林中常見的闊葉樹，與大冇榕極為相似，二者的區別在於豬母乳葉基部較不整正，果實呈圓形；而大冇榕的葉基較植物體及葉形整正，但果實外具有 9-11 個縱稜，另豬母乳的初生葉呈紅褐至黃褐色。低海拔地區將二種植物都稱為大冇樹，”冇”的意思是疏鬆；意思是二種都屬於生長快速且木材疏鬆植物。

攝影／陳子英

豬母乳的花為隱頭花序，當花成熟形成果實時顏色略轉成褐色，並常在樹幹上開花，有樹幹長花的現象。 攝影／陳建志

澀葉榕 *Ficus irisana* Elm.

桑科 Moraceae

澀葉榕又叫糙葉榕，葉的表面粗糙，猶如其名，多分布於低海拔的原始溪谷中，常與大葉楠混生在一起，在較原生的溪谷中也有大型的巨樹，巨樹在基部常分叉開來或在側邊抽出許多萌芽條，但並沒有樹幹長花的現象，在溪谷中常常與菲律賓榕一齊出現，二種葉形極為相似，但澀葉榕與菲律賓榕在葉的基部上，前者葉基較不整正，同時果徑約 0.8~0.9 公分而後者的葉基部較整正，但果徑小於0.5公分。由於葉較粗糙，早期有用它的葉來磨光器物。

攝影／陳建志

楠木屬及樟科植物

香楠*Machilus zuihoensis* Hayata var. *zuihoensis*

樟科 *Lauraceae*

香楠的葉揉而聞之有「載奧辛」的味道，常被人稱為「電線走火味」，木材淡紅色，質地緻密而且輕軟，略具芳香味，早期低海拔地區所稱的「楠木」主要有香楠、紅楠及大葉楠，它們的木材都可以做為建築、傢俱及船舶的材料，樹皮所磨的粉通常叫做楠仔粉，可做為製造線香的高級原料。

攝影／陳子英

大葉楠 *Machilus japonica* Sieb. & Zucc. var. *kusanoi* (Hayata) Liao

樟科 *Lauraceae*

廣泛分布在溪谷或較潮濕的山脊上，也是主要的「楠木」用材之一，木材材質淡紅褐色，適合長期保存，早期都做傢俱、房屋的棟樑或做成雕刻品、木板等用途。在適合生長的森林中常可看見幼苗大量發生，但幼苗死亡率也很高，在生殖更新的策略上似乎採取「人海戰術」。

攝影／曾彥學

豬腳楠（紅楠）*Machilus thunbergii* Sieb & Zucc.

樟科 *Lauraceae*

豬腳楠的分布較特殊主要分布在中海拔闊葉林中，但在台灣北部的陽明山、宜蘭或離島的龜山島、綠島及蘭嶼卻降到低海拔闊葉林中，尤其在衝風的稜線上形成當地森林的主要優勢種。豬腳楠的幼葉和苞片在展開時為紅褐色，配合剛開的黃綠色花朵，常使人有花團錦簇的感覺，在北部較潮濕的亞熱帶地區與常綠闊葉林之墨綠色背景相搭配顯的格外突兀，因此最近常被開發做為庭園樹來栽植。它的木材帶褐色，硬度適中，也是做傢俱、箱櫃及雕刻的主要材料之一。

攝影／陳子英

樟樹 *Cinnamoum camphora* (L.) Presl

樟科 *Lauraceae*

樟樹的葉子有三出脈，樹幹有明顯縱裂，全株都具有樟腦的香味，心材黃褐色至紅褐色，幼樹時材質輕軟，老樹時漸漸變成堅硬，木材常做成雕刻品，如三義的木雕許多就是樟木原料做成的，同時也常被做成傢俱、箱櫃及農具品，但最常被使用的還是做成樟腦油或樟腦膏，早期台灣的樟腦產量一度位居世界之冠，但由於一百多年來的伐採，使得直徑1公尺以上的巨樹已近消失殆盡，只留下如南投和社神木村之樟樹巨木或鄉間的樟樹老樹供後人憑吊，目前各地所看見的多屬於直徑1公尺以下的壯齡木，目前許多鄉間多有栽植樟樹做為行道樹，如集集附近公路的綠色隧道就可欣賞其成行的優美樹形。

攝影／陳子英

殼斗科植物—烏來柯、火燒柯、青剛櫟

烏來柯*Castanopsis uraiana* (Hayata) Kaneh. et Hatus.

殼斗科 *Fagaceae*

分布於北部及中部之低海拔闊葉林，葉上有疏鋸齒，葉基不整正，殼斗外的果實具有瘤狀突起，本樹種砍伐後，常流出大量的樹液，因此又稱為淋漓，在宜蘭及台北烏來地區的溪谷樹形形成傘狀，樹形優美，雖然在森林中屬於冠層突出樹，而且材積在全林分中的比例也極高，但在森林中較少發現幼苗，只有在樹幹旁常常側生萌芽條，同時在野外常可發現這些萌芽條，在老樹枯死後，可以從枯死的老樹幹旁邊向上生長成另一新的植株，這種現象類似竹子由側邊生出新竹一樣。

攝影／陳子英

火燒柯 *Castanopsis fargesii* Fr.

殼斗科 *Fagaceae*

為中部低海拔闊葉林中坡的優勢種，葉背具有茶褐色鱗片，似乎被火燒過，因此稱為火燒柯，果實外的殼斗似小刺狀，葉為全緣或偶有疏鋸齒，木材中心部份呈淡黃褐色，年輪明顯，木材質地堅硬，早期通常做為農具及建築之用。

攝影／曾彥學

青剛櫟 *Cyclobalanopsis glauca* (Thunb. Ex Murray) Oerst var. glauca

殼斗科 *Fagaceae*

分布中及低海拔的森林中，是殼斗科中普遍分布台灣各地的植物，尤其在陡峭的溪邊岩石或東部較乾旱的石灰岩上都可見到它的蹤跡。木材顏色呈灰白色至灰褐色，大多使用於建築及製造器具，目前已有許多低海拔地區栽植為庭園樹供觀賞。

攝影／陳子英

人工造林植物

油桐樹 *Aleurites* spp.

大戟科 Euphorbiaceae

油桐樹指早期由中國長江流域及南部引入栽培於台灣的三年桐及千年桐，兩者都具有大形的葉子，在葉的基部有2個腺體，但三年桐的腺體基部不具柄，同時果實外表較光滑，而千年桐的腺體下面有柄，果實外表具有3個縱稜，同時並有許多皺紋。兩者的果實都可榨取桐油，早期做為塗漆工業的原料，木材材質較輕軟，可做傢俱、木屐及火柴桿的原料。

滿樹白花的千年桐。攝影／陳子英

相思樹 *Acacia confusa* Merr.

豆科 Leguminosae

屬於常綠闊葉樹，它的葉子極為特殊，真正的葉子為羽狀複葉，只有在初生的幼苗才可發現，由小樹開始都形成假葉，在5~6月開出黃色到金黃色的頭狀小花，果實為扁平的莢果，種子扁平褐色，不像小實孔雀豆或孔雀豆所長出的種子一般鮮艷，因此種子也不叫相思豆。由於適應乾旱及強風的環境，再加上高生長的速度很快，長出來的木材堅硬，是薪炭材的良好材料，因此早期在台灣低海拔地區被大量造林。

攝影／曾彥學

演替早期植物

山黃麻 *Trema orientalis* (L.) Bl.

榆科 Ulmaceae

葉子質地紙質，基部歪心形，表面散生許多粗毛，是生長迅速的喬木，通常出現在溪谷旁的崩塌地或開闊地；木材質地輕軟容易腐朽，因此早期常做成木屐、容易腐爛的棺木、火柴桿或做為造紙之原料；也是生長最快速的闊葉樹之一。

攝影／曾彥學

血桐 *Macaranga tanarius* (L.) Muell.-Arg.

大戟科 Euphorbiaceae

也是生長迅速的喬木，通常出現在崩塌地或開闊地，與山黃麻一齊混生在次生的森林中，也可分布至海邊的開闊草叢。葉柄接在葉的中間，形成盾形；木材中心紅褐色，木材質地輕軟，早期做為箱板的材料，嫩葉可做為飼養牛、羊家畜的飼料；是除了竹類植物外生長最快速的闊葉樹之一。

攝影／陳子英

白匏子 *Mallotus paniculatus* (Lam.) Muell.-Arg.

大戟科 Euphorbiaceae

白匏子的葉子基部有一對腺體，葉背密佈白色星狀絨毛，分布在山麓以至海拔1000公尺的山坡次生林或溪谷中；每當微風吹起，常將白色的葉背掀起，就如人微笑時露出牙齒。木材質地非常輕軟，很容易腐朽，早期是製作木屐及薪炭材之材料。

攝影／曾彥學

溪濱植物

茄苳 *Bischofia javanica* Blume

大戟科 Euphorbiaceae

屬於半落葉性的大喬木，當漿質的果實成熟時，成串的果實掛在樹上，格外吸引鳥類來覓食，分布在低海拔的溪谷或沖積平原，由於低海拔地區多已開發為城鎮或其他農業用途，因此雖然茄苳會形成直徑 1~2 公尺以上的巨木，但在野外卻很少見，目前只在低海拔部份地區有保留一些老樹，這些巨木或老樹往往旁邊伴隨著小小廟宇，並常常被人當成神木來膜拜。由於它易於照顧及管理，樹冠寬廣可以遮蔭，同時為誘鳥植物，因此台灣許多街道及公園都常常栽植它做為行道樹或園林樹。

攝影／陳子英

九芎 *Lagerstroemia subcostata* Koehne

千屈菜科 Lythracaeae

落葉的大喬木，它的樹幹極為光滑，但花朵細小，呈白色，多廣泛分布於全台灣地區溪濱或較潮濕的地區，由於莖幹無性繁殖力強，在水土保持上常截取適當大小之莖幹打入溪谷旁之崩塌地，俟其長大可做護坡之用。

攝影／陳子英

櫸（台灣櫸）*Zelkova serrata* (Thunb.) Makino

榆科 Ulmaceae

櫸木的材質很適合做扶梯或地板，並且會在乾季時落葉，在低海拔的闊葉林台灣櫸生長在一年有半乾旱季節的環境，在宜蘭縣的南澳南溪及新竹一帶或台灣中南部及東部地區一年有明顯乾季的地區，就有台灣櫸，但在冬季恒濕的臺北縣烏來和宜蘭縣蘭陽溪則不見它的蹤影。

■櫸(臺灣櫸)在乾季時落葉，生長在半乾旱季節低海拔的闊葉林環境。

攝影／曾彥學

榔榆 *Ulmus parvifolia* Jacq.

榆科 Ulmaceae

盆栽中常用的榆樹指的就是榔榆，榔榆比臺灣櫸生長在更有明顯乾季的地區，尤其在中部地區的河谷中，是它生長較多的地方，它的果實外面具有一圈環翅可以跟台灣櫸果實只具有苞片區分開來，另外葉子的基部也是區分的好方法，榔榆的葉基較台灣櫸偏斜。

攝影／曾彥學

森林中的綠傘

◎撰文／黃曜謀、邱文良
◎ Text ／ Yao-Moan Huang
Wen-Liang Chiou

筆筒樹

Desk Tidy Tree, *Sphaeropteris lepifera*

筆筒樹，隸屬桫欏科，氣生根發達，老葉脫落後會在莖幹上留下橢圓形落葉遺痕。樹高約4公尺或更高，是台灣最大形的樹蕨之一。分布中國南部、琉球、菲律賓，台灣爲其分布中心，一般生長在潮濕又向陽的開闊地，因此台灣北部、東北部、東部及南部等在東北季風影響的區域，全年乾季不明顯，常可看到筆筒樹在這些地區的低海拔開闊山坡地成群生長。在台灣筆筒樹已被廣泛的利用，目前的來源多以山採爲主，應考慮以人工大量繁殖取代山採，確保此一珍貴資源生生不息。

The "desk tidy Tree", *Sphaeropteris lipifera* (Hook.) Tryon, is a member of the Cyatheaceae. These plants have well-developed aerial roots, and oval-shaped scars remain on the trunks after old leaves are shed. It is one of the tallest tree ferns in Taiwan, generally 4m tall or higher. These plants are distributed in southern China, Ryukyus, and the Philippines, and Taiwan is the center of their distribution. Generally, they grow in wet, sunny open areas. Therefore, this fern is a common sight on slopes of low-level open areas of northern, northeastern, eastern and southern Taiwan where is strongly influenced by the northeast monsoon and thus no obviously dry season. The desk tidy tree has already been widely used in Taiwan. Currently, most of these plants are collected from natural areas. We must consider replacing collection with cultivation to protect this precious natural resource and ensure its continued existence.

■混生在闊葉林中的筆筒樹。

攝影／蕨類研究室

筆筒樹(*Sphaeropteris lepifera* (J. Sm.) Tryon) 別稱筆叢樹、粗本筆筒樹、蛇木。隸屬桫欏科，莖約15公分寬，氣生根發達，包圍在莖外圍，葉片長橢圓形，如棕櫚狀叢生於莖幹頂端，長150~200公分，寬70~120公分，三回羽狀深裂，老葉脫落後會在莖幹上留下橢圓形落葉遺痕。一般而言，樹高約4公尺，但是也有高達12公尺以上者，是台灣最大形的樹蕨之一。分布中國南部、琉球、菲律賓，台灣為其分布中心，高大的筆筒樹是台灣產桫欏科中最為人所熟知的種類，產於台灣各地山麓至海拔約1800公尺之濕潤森林中，它一般生長在潮濕又向陽的開闊地，因此台灣北部、東北部、東部及南部等在東北季風影響的區域，全年乾季不明顯，常可看到筆筒樹在這些地區的低海拔開闊山坡地成群生長。

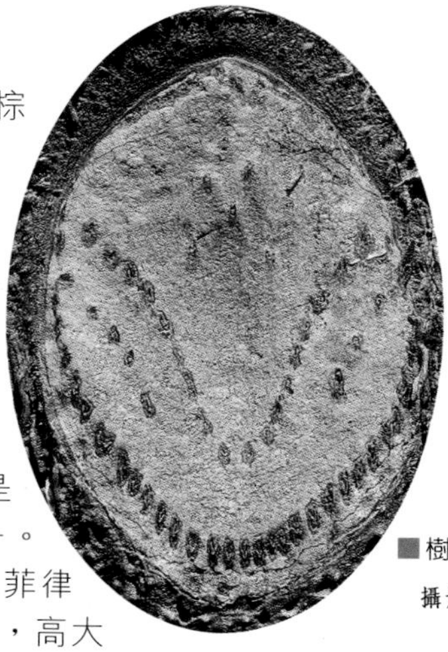

■筆筒樹。 攝影／蕨類研究室

■樹幹上的葉痕。 攝影／蕨類研究室

筆筒樹的生長，根據蘇澳地區資料顯示，唯有當其莖幹高於1.75公尺時，方具生產孢子葉的能力，孢子葉出現的月份從六月到九月，孢子葉抽芽到孢子囊成熟轉褐色所需時間約僅二個月左右，孢

■凹谷地區的筆筒樹林。 攝影／蕨類研究室

子囊多集中在秋季開裂釋放出孢子。一年高度生長量因個體差異頗大，最高達可達45公分，有些則呈停頓狀態。筆筒樹並無如木本開花植物可供判斷年齡之年輪產生，因此要如何知道年輪的筆筒樹有多少歲數呢？有蕨類學者提出，這可以去算算筆筒樹上有多少落葉遺痕，再從它一年會掉落多少片葉子去推算。

■筆筒樹的葉芽。
攝影／蕨類研究室

筆筒樹末裂片腹面著生兩排如蟲卵狀之孢子囊群，每個孢子囊群由13~78個孢子囊所組成，平均值約51，孢子囊內具有64顆孢子，孢子為四面體型，直徑約50微米。孢子播撒於人工培養土約一個星期之後便開始發芽，初生之配子體為絲狀，經由頂端之分生細胞之分裂而成匙形，此時分生細胞漸由分生組織所取代，一個月之後心臟形配子體形成並在腹面左右之翼長出藏精器，心臟形配子體初為單層細胞，而在中間部份發育出多層細胞，稱為中肋，中肋之上半部長出多細胞毛或鱗片，這是桫欏科配子體的重要鑑定特徵。藏精器在中肋尚未發生前即已產生，藏卵器則在第七星期中肋形成後始發生在腹面靠近分生組織之中肋上，此時藏精器仍持續發生，換言之，配子體之性別表現是由雄性至兩性。約在孢子撒播10星期後，藏精器與藏卵器均已達成熟狀，此時澆水，藏精器迅速打開蓋子並釋放精子。根據電泳及配子體的培養試驗證實，筆筒樹傾向於不同配子體間交配，換言之，精子會藉由水分當媒介游到另一配子體的藏卵器中與其卵子結合，這種交配方式可增加族群遺傳變異，促使後代具有更大的競爭能力，結合子經過成長發育約二星期後即可產生幼孢子體。初生數片孢子體葉片為無中肋型，長度約僅1公分左右，葉片上著生許多單列多細胞毛，孢子體隨後長出形態漸趨複雜之中肋型葉子。

在台灣，筆筒樹已被廣泛的利用，像是蘭嶼雅美族人以其嫩芽充當救荒食物，甚至在海面作業時充飢用，其頂部打碎後的黏稠汁液也被應用為洗髮劑。園藝用的蛇木盆、蛇木柱、蛇木板、蛇木屑等，主要即是利用筆筒樹糾纏交結的氣生根製作而成；近年來以其樹姿優雅，被引為庭園木，並廣受青睞。

樹木狀的蕨類在許多國家都將其列為保育類植物，例如中國大陸即將筆筒樹列入紅皮書上的稀有種類。目前在台灣筆筒樹

■傘狀的筆筒樹冠。　攝影／蕨類研究室

■加工中的筆筒樹幹。　攝影／蕨類研究室

利用的來源多以山採為主，這對此種自然資源之保育與永續利用將漸形成威脅。其實筆筒樹每年均可生產不勝其數之孢子，有關其物候學、孢子收集、存藏及繁殖方式也都已有調查及研究，應可考慮以人工大量繁殖取代山採，確保此一珍貴資源生生不息。

■筆筒樹橫切面的花紋。

攝影／蕨類研究室

■樹幹表面的氣生根，黃色是當年剛長出來的。

攝影／蕨類研究室

不可能的任務

石灰岩區植物

The Limestone Plants

◎撰文／呂勝由

◎ Text ／ Sheng-You Lu

當您駐足太魯閣峽谷，仰望那高聳入雲、絕頂危崖，一定很難想像，在人類難以登臨，其他生物大多棄之不顧之處，還會有許多美麗的綠色子嗣俯仰其間，千萬年來，代代爲原本荒漠的地角，塗抹出生命的色彩與律動。太魯閣國家公園中，從蘇花公路海拔零開始到南湖東南峰海拔3,000多公尺，陸續都有石灰岩的出露，百變誘人的植物也隨著海拔的上升而處處驚豔。在太魯閣國家公園的萬紫千紅裡，就屬這岩生植物最引人讚嘆與好奇了，然而到底什麼是岩生植物？什麼是石灰岩植物？在它們無言的姿態裡，又透露了什麼訊息？爲什麼大千世界，它們偏要選這種地方落地生根、世代相傳呢？

When you look at Taroko Gorge, and cast your eyes on the cliffsides rubbing shoulders with the clouds, looking at the dizzy precipices, you will certainly find it hard to imagine that in a place where men find it difficult to gain a foothold, and which are largely eschewed by other forms of life, there are still many beautiful tones of green to meet the eye. For thousands of years, generation after generation, these plants have painted the area with their vibrant color and rhythm. In Taroko National Park the Suhua Highway begins at sea level and moves up some three thousand odd meters to Tungnan. All this distance is limestone, with a variety of plant life changing with the changing altitude. In the multicolored environment of Taroko National Park, these limestone plants are some of the most curious things to be seen, but just what are these plants? What are limestone plants? What is the message sent by their wordless postures? Why, with all the world to choose from, would these plants choose to put down roots and pass their generations here?

■清水斷崖是蘇花公路必經之路，依山面海，景色優美，名聞中外。

攝影／呂勝由

■慈母橋附近Ｖ型峽谷。千萬年來由立霧溪行經石灰岩地區切割形成的Ｖ型峽谷。攝影／呂勝由

台灣石灰岩區分布

台灣是新生代後期隆起的島嶼，位於歐亞大陸的東緣。在淺海地區生活的珊瑚蟲，含高量鈣質的珊瑚遺骸提供一部分造山的素材。由於地殼的擠壓，經高溫高壓作用下，珊瑚礁石更進一步形成變質石灰岩。千萬年來持續不斷的造山運動，使台灣的陸塊不斷抬升。由於變質石灰岩比起板岩、頁岩、砂岩等，質地堅硬緻密，劈理不明顯，不易崩落，在河流的切割作用下，便形成憾動人心高聳入雲的U形峽谷，這就是舉世聞名的太魯閣峽谷的由來。在太魯閣國家公園範圍中，石灰岩出露頻繁，如三棧溪、清水斷崖、太魯閣峽谷、嵐山、金馬隧道、立霧主山、錐麓斷崖、三角錐山、清水山、馱獼陀山、南湖東南峰等，垂直含括3,000多公尺的海拔範圍。在太魯閣國家公園石灰岩地理環境中孕育出其獨特的植物相，由於氣候溫暖潤濕，植物種類繁多，總共約有600種維管束植物，其中約60種屬於台灣稀有植物如清水圓柏（*Junipeurus chinensis* L. var. *tsukusiensis* Masamune）、太魯閣黃楊（*Buxus microphylla* subsp. *sinica* var. *tarokoensis* Lu et Yang）、清水小蘗（*Berberie chingshuiensis* Shimizu）、太魯閣小蘗（*Berberie torokoensis* Lu & Yang）、太魯閣胡頹子（*Elaeagnue tarokoensis* Lu & Yang）等。整體而言，太魯閣石灰岩植群與分布在中國大陸雲南、廣西、貴州西南諸省石灰岩之常綠、闊葉混交林比較，科屬之組成頗為相似。

■錐鹿古道是合歡越嶺古道最令人驚心動魄的一段。 攝影／呂勝由

石灰岩植物之特性

就高等植物而言，岩生植物是生長在岩石裂隙、岩堆或岩屑、土壤稀少地區的植物，一般而言

■清水山石灰岩區孕育許多本省珍稀的植物。 攝影／呂勝由

■錐鹿斷崖。拔昇一千多公尺的崖壁令人驚嘆不已。 攝影／呂勝由

■海拔3,528公尺的南湖東南峰石灰岩區崢嶸嶙峋，令人望而生畏。 攝影／呂勝由

■登三角錐山途中遙望錐鹿山（海拔1,667公尺）的壯麗景致，右上方為台灣五葉松及台灣二葉松點綴其中。 攝影／呂勝由

植物體為適應此惡劣的環境，通常都具有比較發達的根系以利水分及養分的吸收。顧名思義，石灰岩植物是指生長在石灰岩地區的一群植物。典型的鈣土因為土壤中鹼基含量較多，因此呈鹼性。一般而言，近中性的土壤比較適合多數的植物生長，土壤酸鹼度過大對於植物養分的吸收及新陳代謝皆有不利的影響。有一些植物可以適應鹼性土壤的特性，即所謂喜鈣植物或喜石灰岩植物。在雨量稀少的地區，這類植物的優勢度比較明顯；台灣地區的雨量豐沛、淋溶作用旺盛，帶走鈣土中的鈣離子，加上腐植質的累積作用，因此石灰岩土壤乃呈中偏酸性反應，因此要明確的區別喜鈣植物與非喜鈣植物相當困難。生長於石灰岩地區的植物，筆者認為一部份的原因是由於其在一般土壤深厚生育地無法與其它植物競爭，故數量極少，只有在大多數植物不適生長之岩石地，競爭壓力低，可忍受貧瘠環境，便有其一席生長之地。在此，將喜石灰岩植物定義為：喜生育於石灰岩地區之植物，在相似的生育環境下，其生育於石灰岩地區之概率，顯著高於生於非石灰岩地區之概率。然而從廣泛的採集調查中，我們知道有些植物只分布在石灰岩地區，例如：清水圓柏、清水金絲桃（*Hypericum nakamurai* (Masamune) Robson）、細葉蚊母樹（*Distylium gracile* Nakai）、太魯閣木藍（*Indigofera ramulosissima* Hosok.）、清水小蘗、太魯閣黃楊、太魯閣小蘗、太魯閣大戟（*Euphorbia tarokoensis* Hayata）、太魯閣小米草（*Euphrasia tarokoensis* Hayata）、森氏菊（*Chrysanthemum morii* Hayata）等，可做為台灣東部花蓮山地石灰岩地區的指標植物。

石灰岩植物地理

太魯閣峽谷的石灰岩植群與分布在中國大陸雲南、廣西、貴州西南諸省石灰岩之常綠、落葉闊葉混交林比較，科屬之組成頗為相似。兩者除少數之屬如青檀屬、山拐棗屬、魚骨木屬、南天竹屬不產於台灣外，其餘主要之科屬兩地大多相同。其落葉層片主要有榆科、樺木科、漆樹科、胡桃樹科、無患子科、槭樹科及水團花屬、衛矛屬、南蛇藤屬、柿樹屬、薄姜木屬、山菜豆屬、合歡屬。常綠層片主要由殼斗科、冬青科、樟科、木犀科、薔薇科、大風子科、海桐科、茜草科、芸香科、八角楓科、桑科、忍冬科、大戟科、百合科、紫金牛科、瑞香科、虎皮楠科。除上述之科屬相同之外，尚有些如芸香科的黃皮屬、烏柑屬；豆科的皂莢屬、大戟科的巴豆屬、無患子科的欒樹屬等，雖不產於太魯閣石灰岩，但分布於南部之恆春珊瑚礁石灰岩地區；而榆科的糙葉樹屬不產於太魯閣但普遍分布於本省低海拔河岸地區。

在中國的植被類型中「石灰岩常綠、落葉闊葉混交林」為亞熱帶石灰岩山地的原生植物群落，組成種類與山地常綠、落葉闊葉混交林有所區別，以含有榆科及其他喜鈣樹種為特徵。一般皆分布於丘陵及低、中山上。在全中國境內的東部亞熱帶地區，大多分布於海拔1,000公尺以下，在西部亞熱帶地區，則可達海拔1,800~2,000公尺。其主要分布於貴州、廣西及雲南等地區。所在地的石灰岩多露出地面，土壤一般偏鹼性，大多填嵌在石縫中，在山麓才有土壤覆蓋。山麓以上的林木，大多扎根在岩縫中，根系沿著裂縫延伸，一直接觸到有土壤的地方，雖然林木著生受到限制，但林冠一般仍是連續的。

花蓮、太魯閣地區之石灰岩環境

太魯閣國家公園石灰岩區主要的各種植物社會，在非石灰岩地區均可見到，如草叢、灌叢植物社會中之台灣蘆竹型植物社會為本省中、低海拔岩壁極普遍常見之植物社會，而刺柏型植物社會則在中、高海拔之岩壁極為普遍，低海拔河谷之榕楠森林、中海拔之樟櫟林及鐵杉、紅檜林其組成與非石灰岩地區並無二致，以鐵椆或捂桐為優勢之植物社會在本省多石或岩壁之生育地常可見到，以太魯閣櫟為優勢之植物社會在本省其他地區較為罕見，然在非石灰岩地區亦可發現；因此，以植物社會而言，並未有特殊石灰岩生之植物社會。

■錐鹿斷崖。沿著古道的路跡，沿途可見許多岩生植物，其中太魯閣木藍僅見於此，相當稀少。 攝影／呂勝由

花蓮山地石灰岩植物展現其十足的生命韌性，以適應岩壁、石堆的特殊生態環境。基本上屬於貧瘠類型，這類的喬灌木植物根系都比較發達，常能沿石隙伸展並分泌酸類溶解基岩。而草本植物的適應，受岩壁的濕度及溫度影響很大，這也是蕨類植物較少的原因，例如：樹蕨類的配子體生活期至少需要二個月的時間才足夠成為孢子體，因此樹蕨類在本石灰岩區的分布極稀，原因在此。這一片石灰岩植物看似平凡，又有誰真正了解其生存的真義及透露出什麼訊息呢？

花蓮山地石灰岩指標植物

銀杏葉鐵角蕨 *Splenium ruta-muraria* L.

鐵角蕨科 Aspleniaceae

鐵角蕨科之草本植物。分布北美、歐洲、喜馬拉雅、巴基斯坦、日本等地。分布清水山、三角錐山、研海林道等石灰岩地區。

太魯閣小蘗 *Berberis tarokoensis* Lu & Yang

小蘗科 Berberidaceae

小蘗科之常綠灌木。台灣特有種。僅見於太魯閣國家公園三角錐山、研海林道等石灰岩地區。

梓木草 *Buglossoids zollingeri* (A. DC.) Johnston

紫草科 Boraginaceae

紫草科之多年生草本植物。分布中國、韓國、日本。本種於本省僅日人 Fukuyama 曾經採過標本。於太魯閣國家公園分布三角錐山、研海林道石灰岩地區。

太魯閣黃楊

Buxus microphylla subsp. *sinica* var. *tarokoensis* Lu & Yang

黃楊科 Buxaceae

黃楊科之常綠匍匐狀灌木。台灣特產，僅見於太魯閣國家公園三角錐山海拔 1,800 公尺左右之石灰岩區。

森氏菊 *Chrysanthemum morii* Hayata

菊科 Compositae

菊科之草本植物。台灣特產種。分布清水山、三角錐山、錐麓斷崖、綠水、研海林道等地。

太魯閣胡頹子 *Elaeagnus tarokoensis* Lu & Yang

胡頹子科 Elaeagnaceae

胡頹子科之直立灌木。台灣特產。僅見於大魯閣國家公園，九曲洞至天祥一帶之石灰岩及非石灰岩地區。

攝影／呂勝由

太魯閣小米草 *Euphrasia tarokoensis* Hayata

玄參科 Scrophulariaceae

玄參科之多年生草本植物。台灣之特有種。分布三角錐山、研海林道等石灰岩地區。

太魯閣大戟 *Euphorbia tarokoensis* Hayata

大戟科 Euphorbiaceae

大戟科之草本植物。台灣特產種。分布三角錐山、錐麓斷崖、研海林道等石灰岩地區。

清水金絲桃 *Hypericum nakamurai* (Masamune) Robson

金絲桃科 Guttiferae

金絲桃科之常綠灌木。台灣特有種。分布二子山、清水山、三角錐山、嵐山等石灰岩地區。

太魯閣木藍 *Indigofera ramulosissims* Hosok.

豆科 Leguminosae

豆科之常綠灌木。台灣特有種。分布錐麓古道石灰岩地區。

清水圓柏 *Juniperus chinensis* L. var. *tsukusiensis* Masamune

柏科 Cupressaceae

柏科之常綠匍匐狀或直立灌木。台灣特有變種。台灣分布清水山、嵐山及立霧主山之石灰岩地區。

榿葉花楸 *Sorbus alnifolia* (Sieb. & Zucc.) Koch

薔薇科 Rosaceae

薔薇科之落葉小喬木。分布中國、韓國、日本。台灣分布清水山海拔約2,200m之石灰岩地區。

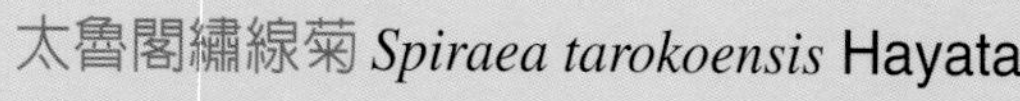

太魯閣繡線菊 *Spiraea tarokoensis* Hayata

薔薇科 Rosaceae

薔薇科之落葉灌木。台灣特產種。分布台灣西部孫海林道、奧萬大等石灰岩地區，及小清水、匯源、研海林道等石灰岩地區。

攝影／呂勝由

魅力四射

秋海棠科植物

The Begoniaceae

◎撰文／彭鏡毅

◎ Text ／ Ching-I Peng

秋海棠科植物在台灣以中低海拔地區為分布的主要範圍，一般說來，台灣原生的秋海棠普遍生長於較為潮濕的土壤或岩壁上，因此在低海拔山區，某些種類的秋海棠植物常可作為潮濕環境的指標物種。在台灣，幾乎全島中低海拔都可以發現秋海棠的蹤跡，而各個台灣原生秋海棠植物物種之水平地理分布與台灣的自然地理分區有著極為密切的關連，其中數種秋海棠呈現廣泛分布，而特有種高達70%以上，台灣原生秋海棠植物可謂精彩絕倫。

The Begoniaceae are primarily found in the mid and low-altitude regions of Taiwan. Generally speaking, Begonias prefer moist soil or rocky slopes.Thus, in low-altitude areas where people hike, a number of Begoniaceae species are often indicative of wet environments. In Taiwan, Begoniaceae are found in practically all the mid and low-altitude regions, and the distribution of these plants in terms of altitude is very tightly linked with the natural biomes in Taiwan. Moreover, continuous distribution is seen in all areas, with some species being very broadly distributed, making studies of Taiwan's Begoniaceae endlessly interesting.

■水鴨腳花朵盛開時以多變的白色、粉紅色與紅色，為婀娜多姿的點點繁花上色，鑲在一片綠意之中更顯耀眼。

攝影／彭鏡毅

■水鴨腳是台灣秋海棠屬植物分布最廣、最為常見的種類之一，在全島中低海拔地區潮濕的林下、路旁邊坡或小山溝常形成壓倒性的優勢分布。

攝影／彭鏡毅

水鴨腳（*Begonia formosana* (Hayata) Masam.）是台灣秋海棠屬植物分布最廣、最為常見的種類之一，在台灣全島中低海拔地區潮濕的林下、路旁邊坡或小山溝常常形成壓倒性的優勢分布，展現以翠綠為大地打底，並不時隨意地染上白色斑紋的風情。花朵盛開時更以多變的白色、粉紅色與紅色為婀娜多姿的點點繁花上色，鑲在一片綠意之中更顯耀眼。水鴨腳因葉型酷似鴨子腳而得名，莖及葉柄肉質、多汁並常泛紅色，鮮嫩的葉柄更為台灣人兒時入口酸甜的回憶。水鴨腳的外部形態相當多變，不論在葉形、葉緣缺刻、葉面色澤、植株被毛、花被顏色與花瓣數目各項特徵上皆有極高的多型性。根據近年筆者等對於水鴨腳親緣地理關係初步的研究顯示，水鴨腳可能是經由天然雜交後，染色體多倍體化而生成的植物，至今並仍持續與台灣其他多種秋海棠進行天然雜交，衍生新種。

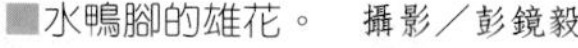

■水鴨腳的雄花。 攝影／彭鏡毅

上■水鴨腳因葉型酷似鴨子腳而得名。

下■水鴨腳的雌花。

攝影／彭鏡毅

最近筆者等發表的一個新種，命名為台北秋海棠（*Begonia taipeiensis* C. I Peng），它是水鴨腳與圓果秋海棠（*Begonia aptera* Blume）天然雜交的後裔；多項特徵介於水鴨腳與圓果秋海棠之間，目前僅發現於台北縣汐止與烏來地區海拔200~500公尺的潮濕山區林道旁與岩壁。由於水鴨腳與圓果秋海棠都廣泛分布於台灣北部的山

上■台北秋海棠是水鴨腳與圓果秋海棠天然雜交的後裔，目前僅發現於台北縣汐止與烏來地區海拔200至500公尺的潮濕山區林道旁與岩壁。

次一■圓果秋海棠。

次二■圓果秋海棠雌花側面觀。

下■裂葉秋海棠的花（左雌右雄）和果實。　攝影／彭鏡毅

■在台灣的低海拔地區有一種著生於山徑的岩壁上，而延伸出數條走莖，再以拍性繁殖產生成新的植株，這種秋海棠是岩生秋海棠。　攝影／彭鏡毅

■岩生秋海棠的雌花。　攝影／彭鏡毅

區，而且生育環境重疊情形相當普遍，因此若透過更進一步的野外調查與採集，或將會發現更多的台北秋海棠族群。除了與圓果秋海棠有天然雜交的情形之外，我們發現水鴨腳可能與裂葉秋海棠（*Begonia palmata* D. Don）以及若干其他種類秋海棠也有天然雜交的情況，尚待更進一步的研究。

■岩生秋海棠的雄花。　攝影／彭鏡毅

在台灣的低海拔地區有一種生活型較為特殊的秋海棠植物，這種秋海棠有著小巧可愛的圓形球莖，常著生於山徑的岩壁上，並延伸出數條走莖，以行無性繁殖產生成新的植株，這種秋海棠叫做岩生秋海棠（*Begonia ravenii* Peng & Chen）。岩生秋海棠的莖葉於秋冬時節結果後逐漸凋落，而在來年的春季再抽新芽，四、五月的開花期間可以見到其雌、雄花各張著兩片粉紅色的花瓣在步道岩壁上向遊人招手，好似造型特殊的風鈴般懸掛山壁上，甚具觀賞價值。岩生秋海棠是台灣的特有種，零星分布於台灣中部苗栗、台中、彰化、嘉義、南投、台南等縣海拔約350~1,000公尺左右山區，數量並不多。

左■每年五、六月盛花時期的蘭嶼秋海棠在蘭嶼全島的道路旁、岩壁上、溪谷中、海邊開闊地珊瑚礁岩上甚至在水泥建築的岩縫中都可以見到它搖曳生姿的身影。

上■生活在石灰岩地的蘭嶼秋海棠。

下■綻放雄花並已結實的蘭嶼秋海棠。　攝影／彭鏡毅

蘭嶼秋海棠（*Begonia fenicis* Merr.）分布於琉球南部島嶼、台灣的蘭嶼及綠島，以及菲律賓北部的巴丹島，其中以在蘭嶼的分布為最為普遍，每年五、六月的盛花期在蘭嶼全島的道路旁、岩壁上、溪谷中、海邊開闊地、珊瑚礁岩上甚至在水泥建築的岩縫中都可以見到它搖曳生姿的身影。蘭嶼秋海棠擁有厚肉質光滑的斜卵形葉片，以不同的方向貼近地面或岩壁上，此時花序上掛滿白或粉紅色的小鈴鐺，好似掛在崢嶸岩壁上的一股溫柔，緩和了炎熱的海風。蘭嶼秋海棠在早期蘭嶼人的生活中是一種相當普遍的食用植物，但他們僅食用嫩葉及莖部，並認為若不慎食用花或花序會罹患疾病。這是蘭嶼秋海棠在民俗食用植物的壓力下仍能開花結果、維持繁茂的原因之一，由此可見蘭嶼達悟族人與大地共生，對生活環境的潛在智慧。

愈捲愈美麗

蕨類植物

Ferns

◎撰文／張藝翰
◎Text ／Yih-Hann Chang

在台灣幾乎任何地方都可以看到蕨類植物，從海岸地區到山坡地，從低海拔到高海拔，從森林到鄉村、都市，不論在哪個林帶、什麼環境，都有蕨類的縱跡。而蕨類植物在適應這些千變萬化的環境，也各有一套不同的生存策略。

Ferns can be seen literally everywhere in Taiwan. They occur from the coastal areas to the high mountains, from the low altitudes through the highest areas. They are seen from the forests to the villages, and even in the cities. Regardless of the place or environment, the traces of ferns are always seen. In adapting to these wildly varied environments, ferns have themselves adopted a wide variety of strategies for survival.

■烏毛蕨的幼葉。

攝影／蕨類研究室

低海拔海岸植群

海岸附近地區由於存在著陽光強、溫度高、海風凜及水分少等惡劣生長條件，限制了蕨類植物的生育與發展，因此能夠征服這種惡劣環境的蕨類勇者並不多。為了克服不良環境，生存在此間的蕨類植物大多具有植株縮小化、葉片厚革化、根系深廣化、耐高溫化、耐生理乾旱化...等適應環境的機制。當這些海邊蕨類一旦逐漸遠離海岸時，則會有形體逐漸變大且葉片趨於轉薄的趨勢。

鹵蕨 *Acrostichum aureum* L.

在南洋及美洲地區的紅樹林附近，常可見到本種之族群。但在台灣的鹵蕨則分別分布於南部海邊的淡水出海口及東部海岸山脈泥火山的地區，數量非常稀少。

全緣貫衆蕨 *Cyrtomium falcatum* (L. f.) Presl

常見於海濱附近的山坡或礁岩罅縫處。

傅氏鳳尾蕨 *Pteris fauriei* Hieron.

在海濱附近的山坡或礁岩罅縫處非常常見。

闊片烏蕨 *Sphenomeris biflora* (Kaulf.) Tagawa

在海濱附近的山坡或礁岩罅縫處常見。

海岸擬茀蕨 *Phymatosorus scolopendria* (Burm.) Pichi-Sermolli

在海濱附近的珊瑚礁岩或岩壁上時常出現。

■海岸擬茀蕨。

攝影／蕨類研究室

■鹵蕨。

攝影／蕨類研究室

低海拔熱帶森林（榕楠林）

熱帶蕨類的特色主要為根莖匍匐橫走、葉片大型且薄、葉片分裂程度較低且常為不分裂的單葉、葉片邊緣多為全緣而少鋸齒、常有兩形葉（dimorphic leaves/fronds）的分化、無性繁殖現象頗為常見。

網脈突齒蕨 *Pleocnemia submembranacea* (Hayata) Tagawa et K. Iwats.

屬於泛全台低海拔分布之種類，通常在受干擾程度較低而且濕度適中的闊葉林下，常呈高密度的群生。本種在台灣中部地區出現之頻率相當高，但在南部及北部則稍有降低。

沙皮蕨 *Hemigramma decurrens* (Hook.) Copel.

屬於泛全台低海拔分布之種類，不過它一般較常出現在受干擾程度較高、濕度偏低而且有適度遮陰的林下斜坡或道路兩側。

南洋巢蕨 *Neottopteris australasica* J. Sm.

巢蕨又叫鳥巢蕨或山蘇花，是一種常見於樹上的大型著生植物。在台灣的巢蕨類中，許多人都曉得有山蘇花（*Neottopteris antiqua* (Makino) Masam.）和台灣山蘇花（*Neottopteris nidus* (L.) J. Sm.）兩種，但知道南洋巢蕨也分布於台灣而且為數不少的人就不多見了。其實南洋巢蕨是台灣三種巢蕨中分布海拔最低、最靠近海岸的種類，聽說也是最好吃的一種。

藤蕨 *Arthropteris palisotii* (Desv.) Alston

全台低海拔山區分布的著生植物，愈往中南部愈是常見，通常會沿著岩壁或樹幹成群狀或片狀著生，甚至會盤據整個大樹樹幹或岩壁的表面。

羅蔓藤蕨 *Lomariopsis spectabilis* (Kze.) Mett.

是一種稀有的中大型低位著生植物，主要盤據於大樹樹幹或巨石的基部附近。本種在台灣之主要分布區域為南部及東南部濕熱的低海拔山區，不過在台北和宜蘭境內也有十分零散的族群。

■沙皮蕨。　攝影／蕨類研究室

■藤蕨。　攝影／蕨類研究室

■山蘇。

攝影／蕨類研究室

低海拔亞熱帶森林（樟楠林）

從蕨類植物來看，分裂程度較高且葉片邊緣具有鋸齒的中、大型種類逐漸增多，常成為地面主要優勢種（dominant species）；而此間分布的蕨類在比例上，與中國大陸華南地區的親緣及地緣關係較為密切。而熱帶森林林內之蕨類，雖然也常出現在本區範圍，但族群數量已明顯變少。

廣葉鋸齒雙蓋蕨 *Diplazium dilatatum Blume*

本種是全省低海拔濕度適中之森林下非常優勢的中大型草本植物，不僅經常於林下地表形成廣闊而綿密的優勢族群，而且又因具有較大的生態幅度（ecological amplitude），即使在中海拔相當潮濕的暖溫帶林下或是靠海且乾燥的山坡植群中，也經常可以見到它的蹤跡，只不過其優勢地位已被其它種類取代。

■廣葉鋸齒雙蓋蕨。 攝影／蕨類研究室

過溝菜蕨 *Diplazium esculentum* (Retz.) Sw.

本種喜濕喜光，在低海拔向陽或略遮陰的溝渠兩側及河川溪谷的礫石地上，經常可見到它的繁盛族群。過溝菜蕨由於其嫩芽經炒食後非常美味可口，故早已為人推廣成蔬菜作物。

■雲南三叉蕨。 攝影／蕨類研究室

雲南三叉蕨 *Tectaria yunnanensis* (Bak.) Ching

大型草本，喜濕喜陰，主要分布在全省各地潮濕溪谷附近的山坡或石縫處，尤以北部、東北部及東南部為多。

筆筒樹*Cyathea lepifera* (J. Sm. ex Hook.) Copel.

全省各地常見的喬木狀樹蕨類，也是台灣產蕨類中最大型者，其高可達20公尺以上。雖然說是全省性分布，但由於桫欏科植物的配子體有不能忍受數月乾旱的特性，所以筆筒樹的分布仍以北部及東北部之低海拔山區為多；特別是宜蘭羅東至蘇澳一帶，有其最豐沛的族群。

■筆筒樹。 攝影／桂曉芬

崖薑蕨*Aglaomorpha coronans* (Wall. ex Hook.) Copel.

全省低海拔地區林木樹幹上著生的大型蕨類，尤其是在空氣濕度較高的原生林內，出現的程度更加頻繁。

中海拔森林（樟殼林及檜木林）

就蕨類的觀點而言，在中海拔的蕨類植物，不僅物種多樣性急速增加，而且在種類組成上亦以鱗毛蕨科（Dryopteridaceae）和蹄蓋蕨科（Athyriaceae）為大宗，至於著生植物方面，則是由水龍骨科（Polypodiaceae）的蕨類植物大放異彩。而在形態特徵方面，具有直立性塊狀根莖、葉片薄、葉片分裂的回數多、葉緣多鋸齒的蕨類佔了相當大的比例。

稀子蕨 *Monachosorum henryi* Christ

在全省中海拔山區林下極為常見，其最顯明的特徵是在其葉軸上經常有一至多枚外形酷似拳頭狀的不定芽，可行無性的營養繁殖。本種常與鱗毛蕨科的魚鱗蕨（*Acrophorus stipellatus* (Wall.) Moore）相偕出現。

川上氏雙蓋蕨*Diplazium muricatum* (Mett.) v.A. v.R.

全省中海拔山區普遍分布的中型地被蕨類植物，辨識的特徵是在其葉柄基部有許多半透明的刺狀物。環境適宜的話，常獨自或與其它雙蓋蕨類如奄美雙蓋蕨（*Diplazium amamianum* Tagawa）、德氏雙蓋蕨（*Diplazium doederleinii* (Luerss.) Makino）...等共同組成地被優勢草本。

廬山石葦 *Pyrrosia sheareri* (Bak.) Ching

在全省中海拔山區相當常見的著生植物。

■稀子蕨。 攝影／蕨類研究室

台灣瘤足蕨 *Plagiogyria formosana* Nakai

在海拔較高之檜木林或是針闊葉混生林內，其地表是由台灣瘤足蕨單獨或是夾雜著許多不同種類的鱗毛蕨科植物擅場的天下。台灣瘤足蕨最明顯的特徵，就是它具有粉白甚至有點亮白色的葉背，十分容易辨識。另外，由於本種的繁衍生長力十分旺盛，常在林下拓殖出偌大的族群；而當山風吹拂，鮮白色之葉背此起彼落地翻轉搖曳的畫面，不可不謂之為台灣山地一項相當奇特的景觀。

巒大蕨 *Pteridium revolutum* (Bl.) Nakai

在較為乾燥或是曾經發生火災、崩坦、嚴重人為干擾後的中海拔山地以及該海拔範圍內的一些針葉林地，其林相十分單純，地被植物的組成也十分貧瘠，而巒大蕨正是其中最多、最優勢的少數種類之一。

■廬山石葦。

攝影／蕨類研究室

高海拔植群

一般來說，高海拔山區的生存環境具有風強、溫低、霜雪、土壤層淺薄以及土壤養分貧瘠...等特性。所以高海拔蕨類為了適應這些不良生長條件，通常會採取形體縮小、根莖直立、根系深廣、葉片被覆密毛或鱗片、冬季凋萎、能於低溫下進行生理代謝作用、在開闊地上盡量貼伏地表或藏身於岩礫溝縫處...等策略來應付。在組成種類方面，乃是以石松科（Lycopodiaceae）、鱗毛蕨科的鱗毛蕨屬（*Dryopteris* spp.）和耳蕨屬（*Polystichum* spp.）、以及蹄蓋蕨科的蹄蓋蕨屬（*Athyrium* spp.）為主。

冷蕨 *Cystopteris fragilis* (L.) Bernh.

高海拔針葉林下非常普遍的小型蕨類，多為地生，偶見岩生。一般來說，本種喜歡潮濕且遮陰的環境，但是在略為向陽的路邊岩礫縫處，也偶爾可以看到其少數族群。

福山氏耳蕨 *Polystichum wilsoni* Christ

本種在高海拔潮濕的針葉葉林下經常可見，屬於地生的中型蕨類；最大的特徵是其葉片被覆著非常濃密的白毛。常與闊葉鱗毛蕨（*Dryopteris expansa* (Presl) Fraser-Jenkins et Jermy）、逆羽蹄蓋蕨（*Athyrium reflexipinnum* Hayata）、對生蹄蓋蕨（*Athyrium oppositipinnum* Hayata）...等種類混生。

玉柏 *Lycopodium juniperoideum* Sw.

在高山開闊的坡地或稜脊上，常可見到玉柏的族群，常與玉山石松（*Lycopodium veitchii* Christ）、玉山地刷（*Lycopodium yueshanense* Kuo）、石松（*Lycopodium japonicum* Thunb.）...等擬蕨類（fern allies）混生。

杜氏耳蕨 *Polystichum duthiei* (Hope) C. Chr.

鱗毛蕨科耳蕨屬中的小型蕨類，主要分布在高山乾燥且空曠的石礫坡或圈谷附近的岩縫和礫石堆中，族群不算稀少。

線葉鐵角蕨*Asplenium septentrionale* (L.) Hoffm.

稀有的小型蕨類，主要分布在高山稜脊、上坡或圈谷附近的岩縫處。

■玉柏。　攝影／蕨類研究室

■線葉鐵角蕨。　攝影／蕨類研究室

石灰岩地區

台灣主要之變質石灰岩帶，起自蘇花公路和平溪以北的谷風，向南延伸至台東關山以西，總長約150公里；其中發育最佳者，乃是在清水斷崖一帶。由於石灰岩（limestone）基質含有高量的含鈣成分，所以由此發展出的土壤之含鈣量亦高，並呈現明顯的中性或鹼性，與其它地區呈酸性的土壤有很大的差異，形成的植群自然也不同。適合生長於石灰岩帶的植物，多半是能忍受、適應、甚至偏好高含鈣量土壤的好鈣植物（calcicoles）或喜鈣植物（calciphiles）；這類植物除了稀有種（rare species）的比例偏高外，多倍體（polyploidy）及生態品種（ecotype）出現的頻率亦較高。

俄氏鐵角蕨 *Asplenium oldhami* Hance

低海拔石灰岩地區常見的小型蕨類，屬於石灰岩地區的偏好種。族群通常多為岩壁生，地面生的情況也頗常見，但樹幹生的情形就很少看到了。

擬密葉卷柏 *Selaginella stauntoniana* Spring

本種在中、低海拔的石灰岩地區頗為常見，但也可見於全省各山區的岩石壁處。

細葉鐵角蕨 *Asplenium pulcherrimum* (Bak.) Ching

細葉鐵角蕨為一種非常稀有的小型蕨類植物，除了在花東的中海拔石灰岩地形可以找到（低海拔的石灰岩區域也可能出現，但機率很小）之外，其它地區幾乎沒有其分布。

■擬密葉卷柏。 攝影／蕨類研究室

■細葉鐵角蕨。 攝影／蕨類研究室

雲霧繚繞的天堂

中海拔
The Mid-Altitudes

◎撰文／郭城孟 ◎ Text ／ Chen-Meng Kuo

攝影／蕨類研究室

■神祕湖。攝影／蕨類研究室

台灣中海拔地區主要雲霧繚繞的森林，包括海拔較高的以扁柏或紅檜為主的涼溫帶針闊葉混合林生態系，即「檜木林」帶，及較低海拔地區以殼斗科植物為主的暖溫帶闊葉森林生態系，即「樟殼林」帶。

台灣的天然針葉林包括冷杉林、鐵杉林、檜木林，分布在海拔1,800~3,500

Taiwan's mid-altitude regions are primarily cloud-shrouded forests, including the cool temperature conifer mixed forests in the relatively higher altitudes where *Chamaecyparis obtusa* Sieb. and Zucc. var. *formosana* Hayata) or *C. formosensis* Matsum. are the dominant species. This is the cedar belt. Where lower altitudes form a warmer temperature belt of broadleaf forest ecosystem with the oak predominating we see the Lauraceae-Fagaceae forest belt.

■七家灣溪的濱溪森林。

攝影／蕨類研究室

■鴛鴦湖地區的苔林，樹幹佈滿蘚苔植物。 攝影／蕨類研究室

Taiwan's natural conifer forests include fir, hemlock and cedar forests, and are distributed between 1,800 and 3,500 meters altitude. The broadleaf forests of Taiwan run from the flatlands of the lower altitudes to around 2,500 meters altitude. The ceadar forest belt is the line where the broadleaf and conifer species mix, and the canopy is made up of tall conifers -- Formosana red cypress or Taiwan cypress, with the lower layers made up of broadleaf forests, dominated by camphor and oak. These forests have a clear delineation between canopy layer, shrub layer and herb layer, so that the compared to the average conifer forest, the camphor forest is complex and diverse. In this forest belt, relic species are a major characteristic. Taiwan has many types of Tertiary period relic species such as *Taiwania cryptomerioides* Hayata, *Cunninghamia konishii* Hayata and so on. These relic species are mostly distributed in the camphor forest belt or areas above the climate for which reason it can be deduced that after Taiwan emerged from the sea during the Tertiary period, the climate on the island until the coming of the glacier in the Quatornary period may have been similar to that found in the 1800 to 2,500 meter range today. This is a cool, moist climate, and vegetation here includes conifer and broadleaf tree species.

公尺之間，而台灣闊葉林則由低海拔的平地分布至海拔2,500公尺左右的地區，而檜木林帶是一處針葉林與闊葉林的交會帶，樹冠層是高大的針葉樹一紅檜或扁柏，下層卻有如闊葉森林的結構，主要是以樟科、殼斗科植物為主的闊葉林，具有喬木層、灌木層、草本層之分化，所以檜木林的結構比起一般針葉樹林，是複雜而多樣化的。在這一個林帶，孑遺植物眾多是重要的特色，台灣有許多第三紀孑遺植物如台灣杉、巒大杉等，這些孑遺植物大都分佈於檜木林帶或氣候上類似的地區，由此推測台灣在第三紀中葉經造山運動浮出水面之後，到第四紀冰河來臨之前，當時的大氣候可能就類似今天海拔1,800~2,500公尺的地區，是一種涼爽、潮濕的氣候情況，而植物群落則包含了針葉樹種與闊葉樹種。

Taiwan's cedar trees are broadly distributed, but a very large area of these trees has already been logged. On Alishan, Taiping Mountain and other areas, the cedar cypress are very long-lived, and so Taiwan's "giant" are often junipers. But it's worth considering that the "giant" we see today were often considered to be plants without economic value during past periods of heavy logging. Compared to their former peers, they have not developed as well, and were spared as a result. Cedar like to grow on landslides areas or very hilly places, and in Taiwan with her relatively young geological age, slopes and tendencies toward landslides are very common. This is why cedar forests are particularly representative of Taiwan's natural environment. Because of the plentiful rainfall in these forests, ponds or swamps often form in the low-lying areas in the saddles between the mountains where the rainwater finds poor drainage, and the location of these within a terrestrial forest forms a very unique aquatic ecosystem. They are like an island amid the sea of trees. There is often a considerable distance between these ponds, and the gene pools of the ponds have little chance to

■霧氣瀰漫的神秘湖，樹幹上長滿了著生植物。　攝影／蕨類研究室

台灣檜木分布極廣，但已被砍伐了很大的面積，像阿里山、太平山等地即是，檜木是很長壽的樹種，所以台灣的「神木」大部分都是檜木。不過值得我們深思的是，今天我們所看到的神木，往往是過去大肆伐木時期被認為沒有經濟價值的植株，因為它們比起當時附近的同伴，可說是發育最不好而被留下來的。檜木喜歡生長在崩塌地、坡度較陡的地方，與台灣年輕的地質年代，陡峭而容易崩塌的山嶺地形可說是極為匹配，所以檜木林是台灣最具代表性的生態環境之表相。檜木林帶由於降水量豐富，在兩山鞍部窪地排水不良的地方易形成山地池沼，由於位處陸域森林之中，形成相當獨特的水生生態系，就像林海中的孤島，池沼彼此之間相距甚遠，物種基因無法自然交流，極易形成獨立演化的情形，所以台灣除了高山地區具有島嶼特性之外，山地池沼的生物種類也頗具地方性色彩，例如鴛鴦湖、神秘湖、松蘿湖等，池沼四週為檜木林所環繞，環境冷涼潮濕，樹幹與樹枝滿佈苔蘚植物，形成所謂的「苔林」。

到了海拔1,800公尺以下檜木會逐漸消失，顯露出下層之闊葉森林，即為「樟殼林」帶，這類植物的葉子比低海拔的亞熱帶闊葉林種類小，葉表常具較厚的角質層，質地較硬。此一林帶的下限在北台灣地區約為海拔500公尺、南部則是700公尺，這是由於台灣南北緯度上的差異所造成的結果。樟殼林的層次結構多可分成四

spread naturally. This means that independent evolution is very common, and so apart from high mountain zones with gorge characteristics, there are also unique localized species living in the high-altitude ponds, such as Yuanyang Lake, Shenmi Lake, Songluo Lake and so on, where the ponds are surrounded by cedar on all sides, and the environment is wet and cool. The tree trunks and tree branches are covered with mosses and lichens, creating the so-called "mossy forest".

The cedar gradually disappear at altitudes of below 1,800 meters, and the lower layer of broadleaf forest appears. This is the oak belt, whose plants have leaves that are smaller than the leaves of the trees in the subtropical broadleaf forest. The surface of these leaves often have a thicker stratum corneum, and are harder. This forest belt's lower limit is around 500 m in northern Taiwan and 700 m in the south. This is because of the effects caused by the difference in latitude in the northern and southern parts of the island. The layer structure of the camphor forest is fourfold, with one or two layers of trees, a shrub layer and a herb layer. The more layers there are, the more rainfall is retained. When it rains, the leaves on the topmost layer of tree branches are the first to catch the water, which then flows past smaller branches and is collected in the larger ones, to move later through the large branches and trunk to enter the roots to be absorbed by the tree. Any rainwater which is not captured by this topmost layer moves down to the second layer of canopy where it is once again trapped, flowing down the branches to the ground and permeating the soil; moreover, there is also a layer of shrub and ground cover that catches water, plus the epiphyte living on trunks and branches, which also absorb a great quantity of water. Therefore, in times of

層，即喬木第一、二層、灌木層與草本層，層次愈多雨水的截留可更完全：降雨時第一喬木層樹枝上的樹葉最先截留雨水，流經小枝條並匯集至較大枝條，再經由支幹、主幹而滲入地下由根部吸收，未被第一層樹冠截留的雨水，至第二喬木層的樹冠再次被截留，順著枝條流到地面，滲入地下；此外尚有灌木層與地被層的截留，再加上生長在樹幹及枝葉上的著生植物也吸取相當大量的水分，所以驟雨時闊葉林的土壤不會受到直接衝擊，雨水有足夠的時間滲入地下，此即森林涵養水源及水土保持的功能。林下土壤也是森林自然演替的種子庫，也可由森林結構而得到保護，所以說天然闊葉林是目前大氣候環境下完美的產物。

台灣暖溫區闊葉林帶的森林，除了層次結構的分化之外，尚見著生植物與木質藤本植物穿插在各個層次間，造成多變化的棲息環境，加上地質、地形、氣候等因素，以及所佔海拔範圍廣大，台灣大約有43%的生物種類生活在此一環境。然而目前台灣的暖溫區闊葉林帶，多被闢成菜園、果園或柳杉人工造林地，而原來此區的主要樹種是殼斗科及樟科植物，是無數野生嚙齒類動物的主要食糧，尤其是殼

rainfall, the soil of the broadleaf forest does not receive much direct impact. The rainwater has sufficient time to permeate downward into the soil, which gives the forest a source of water and the ability to retain soil and water. The soil under the forests is a storehouse of seeds for succession. These can also be protected by the structure of the forest itself. Therefore, it is said that the natural broadleaf forest is currently the most complete product of the temperate environment.

The forests in the warmer areas of Taiwan's broadleaf forests are not only divided into a layered structure, but also have epiphyte and vines distributed in and between their layers. These create more varied habitats. Adding topography, geological and climatic factors to this mix, and the broad area of altitudes encompassed, it results that about 43% of Taiwan's animal and plant species live here. Even so, currently Taiwan's temperate broadleaf forest belt has mostly been cut down to make vegetable farms, orchards or planted forests of *Cryptomeria fortunei*. The original predominant species in these areas, the oak and the camphor species, were the prime food for innumerable wild rodents, particularly the oak. The changes in the face of the broadleaf forests, plus the direct or indirect eradication of the natural enemies of small mammals, such as snakes or hawks, by human activities, has led to a huge increase in rodent populations, coupled with a lack of the food formerly pro-

■赤楊落葉林。

攝影／蕨類研究室

斗科的植物，而在天然闊葉林經林相變更後，加上如蛇類、鷹類等小型哺乳動物天敵直接或間接地為人類捕殺，小型嚙齒類動物大量繁殖，又缺乏天然林提供所需的食物，故轉而啃食樹皮，而首當其衝的自然是柳杉造林地，常見柳杉林枯死的現象，即是遭飛鼠啃食樹皮而將植株環狀剝皮所致，所以中海拔地區人工林鼠害頻繁，其癥結或許就是生態不平衡的緣故。

中海拔森林受到干擾之後，會自然演替形成以赤楊（*Alnus formosana* (Burk.)

vided by the natural forest. These rodents have turned to nibbling on tree bark, and the first victims of their teeth are the Cryptomeria forests. The phenomenon of dying Cryptomeria has been caused by flying squirrels gnawing on the bark of the trees, effectively girdling them. As a result, the cause of the rise in population of these squirrels in planted forests at the planted regions is the result of nothing more or less than an imbalance in the natural ecological system.

After being disturbed, the mid-altitude forests will naturally evolve into a deciduous forest marked by *Alnus formosana* (Burk.) Makino, or Aceraceae species. A. *formosana* (Burk.)

■混生在二葉松林中的落葉樹—楓（*Acer*）。

Makino）或楓樹科（Aceraceae）植物為主的「落葉林」。赤楊也是孑遺植物，喜歡在開闊地生長，是檜木林被破壞的指標。因為赤楊根部有根瘤菌共生，會將氮肥固定，使土壤肥沃，因此原住民會在墾植數年的土地上種植赤楊，讓土地恢復地力。而楓樹科植物即是膾炙人口的紅葉景觀主角，中海拔地區在秋冬之際，會形成滿山的楓紅，其實這是檜木林或暖溫帶闊葉林被破壞的見證。

Makino is another relic species here, which likes to grow in open areas. This is the sign that a cypress forest has been cut down. Because the roots of *Alnus formosana* have fungi on them, they have a nitrogen-fixing function, and enrich the soil. Consequently, the aboriginal peoples of Taiwan planted these trees on land that had been open for many years, to allow the soil to regain its vigor. The Aceraceae species are marked by their fiery red leaves, and in the autumn, in mid-altitude regions, these plants will fill a mountain with their scarlet hues. This is a sign that a cypress forest or temperate broadleaf forest has been logged.

攝影／蕨類研究室

雲霧中的身影

樟殼林

The Mid-Altitude Broadleaf Forest

◎撰文／陳子英
◎Text／Tze-Ying Chen

■北插天山台灣水青岡林。

攝影／陳子英

中海拔的闊葉林分布於北部700到2,000公尺，在中部1,500到2,500公尺，主要由果實具有殼斗的殼斗科（*Fagaceae*）和全株具芳香的樟科（*Lauraceae*）植物所組成，由於林帶中常有雲霧繚繞，因此也稱爲霧林帶，依海拔可分成上下二層，上層混生有針葉樹甚至形成純林，下層則由樟科及殼斗科之大喬木所組成，本帶的樹木較低海拔的闊葉林高大，有到達20~30公尺，森林結構的分層可分成3~4層，樹幹上常附生有蘭科、蕨類及苔蘚類植物，整個森林由於結構的分層，因此物種歧異度極高。在林帶中有部分常見的落葉樹種，如台灣赤楊、台灣胡桃、台灣檫樹、山桐子、槭樹科植物、千金榆屬植物及阿里山榆等，尤其在北部地區的稜線上有稀有的台灣水青岡夏綠林，在瘦脊的稜線上也常有常綠的台灣杜鵑灌叢。日治時代至光復初期，由於木材的需求，栽植有柳杉林及檜木林，目前許多地區已成林，本林帶物種多樣性極爲豐富，希望我們能好好保存供後代子孫使用。

Taiwan's mid-altitude broadleaf forests are distributed between 700 and 2,000 meters in the north, and between 1,500 and 2,500 meters in central Taiwan. These forests are primarily composed of Fagaceae and Lauraceae. Since these forests are often shrouded in fog and cloud, they are also called fog forests. Two layers are distinguishable depending on the altitude. The upper layer is mixed with conifers, or even pure forest, while the lower layers are composed of large trees from the Lauraceae and Fagaceae families. The trees here are larger than those in the low-altitude broadleaf forests, and may reach 20 to 30 meters in height. The forest has three to four layers, and epiphyte orchids, ferns and mosses often attach themselves to the tree trunks. Because of the division of this forest into layers, there is very high diversity of species. There are some commonly seen deciduous species in the forest belt, such as Taiwan alder, Taiwan walnut, Taiwan sassafras, idesia, Acer species, Carpinus cardata and Alishan elm. Particularly in the north, the rare *Fagus hayatae* Palib. *ex* Hayata is seen atop the narrow mountain peaks. Atop the narrow lines of the ridge of the mountains, evergreen Taiwan azalea shrubs can be seen. During the period of Japanese occupation to the early days after Retrocession, because of the demand for wood, Cryptomeria fortunei and cypress forests were planted, and now these plantings have become forests in many places. With the great diversity of plant species in this belt, we hope to conserve them well to pass them on to future generations.

環境概述

樟殼林海拔由1,500到2,500公尺，北部則降至700到2,000公尺，氣候溫和，空氣濕度極高，常有雲霧繚繞，形成雲霧帶，所謂雲霧站在上方觀看即為雲海，在這一帶中可依海拔的高低分成上層及下層，上層在中部地區海拔約2,000~2,500公尺，下層在中部地區海拔約1,500~2,000公尺，主要樹種都由闊葉樹組成，但在上層有較多之針葉樹，部份針葉樹並形成純林，如扁柏林或紅檜林，下層則主要由樟科和殼斗科的大型闊葉樹優勢喬木所組成；在森林結構上發展成樹冠層、中層、下層及地被層，類似都市的大樓或多層公寓，樹冠層主要由殼斗科的狹葉櫟（*Cyclobalanopsis stenophylloides* (Hayata) Kudo & Masamune *ex* Kudo）、大葉石櫟（*Pasania kawakamii* (Hayata) Schottky）、毽子櫟（*Cyclobalanopsis sessilifolia* (Bl.) Schottky）、赤柯（森氏櫟）（*Cyclobalanopsis morii* (Hayata) Schottky）、長尾尖葉櫧（長尾柯）（*Castanopsis cuspidata* (Thunb. *ex* Murray) Schottky var. *carlesii* (Hemsl.) Yamazaki f. *carlesii*）、錐果櫟（*Cyclobalanopsis longinux* (Hayata) Schottky var. *longinux*）等植物及樟科的紅楠、霧社木薑子（*Litsea elongata* (Wall. Ex Nees) Benth. & Hook. f. var. *mushaensis* (Hayata) J.C.Liao）、假長葉楠（*Machilus japonica* Sieb. & Zucc.）、長葉木薑子（南投黃肉楠）（*Litsea acuminata* (Bl.) Kurata）、烏心石（*Michelia compressa* (Maxim.) Sargent）、墨點櫻桃（*Prunus phaeosticta* (Hance) Maxim. var. *phaeosticta*）、香桂（*Cinnamomum subavenium* Miq.）及木薑子屬（*Neolitsea*）等植物所構成，下層則由山茶科（*Theaceae*）的柃木屬、楊桐屬（*Adinandra*）、冬青科（*Aquifoliaceae*）的冬青屬（*Ilex*）植物、灰木科的灰木屬、木犀科及五加科等植物，下層則常見有狹瓣八仙花（*Hydrangea angustipetala* Hay.）、紫金牛科植物所構成，地被常見有蕁麻科植物、稀子蕨(*Monachosorum henryi* Christ）、蛇根草（*Ophiorrhiza*）及複葉耳蕨（*Arachniodes*）、雙蓋蕨（*Diplazium*）、瘤足蕨（*Plagiogyria*）與鳳丫蕨（*Coniogramme*）等蕨類。在樹冠層、中層及下層常有大型藤本盤旋於各層之間，如五味子科（*Schisandraceae*）、鑽地風（*Schizophragma*）、大枝掛繡球（*Hydrangea integrifolia* Hay.*ex* Hay.et Matsum.）、細小的藤本如長果藤（*Aeschynanthus acuminatus* Wall. *ex* A. DC）、柚葉藤（*Pothos chinensis* (Raf.) Merr.）則常附在樹幹上。樹幹上則有許多附生植物

左上■南澳南溪中海拔闊葉林。
右上■思源埡口霧林。
左下■南澳南溪中海拔闊葉林。
右下■沙里仙溪中海拔闊葉林。

攝影／陳子英

■神祕湖霧林。

攝影／陳子英

(epiphyte plant)生活在各層之中，如蘭科植物的石槲蘭屬（*Dendrobium*）、羊耳蒜屬（*Liparis*）、豆蘭屬（*Bulbophllum*）及蕨類植物如山蘇花（*Neottopteris antiqua* (Makino) Masam.）、波氏星蕨（*Microsorium brachylepis* (Barker) Nakaike）、骨碎補屬（*Davallia*）、水龍骨屬（*Polypodium*）、瓦葦屬（Lepisorus）、石葦屬（*Pyrrosia*）、膜蕨科（*Hymenophyllaceae*）植物及小膜蓋蕨（*Araiostegia perdurans* (Christ) Copel.）等。也由於森林結構的分層使得中海拔森林形成植物歧異度最高的林帶。

■小鬼湖附近中海拔闊葉林。

攝影／陳子英

■神秘湖霧林帶。

攝影／陳子英

稜線上的杜鵑

■中海拔之台灣杜鵑。　攝影／陳子英

在中海拔闊葉林或針葉林中700~2,400公尺的狹窄稜線或陡坡處，常生長樹幹分叉且多彎曲的台灣杜鵑樹叢，在中海拔樟殼林下層的闊葉林中多形成純林，在海拔較高以針葉樹為主的扁柏林中則構成森林下層的主要優勢種，整個樹叢的高度通常只有4~5公尺，樹叢上常掛滿苔蘚類植物，下層之地被植物較單純，主要為玉山箭竹（*Yushania niitakayamensis* (Hayata) Keng f.）或瘤足蕨類植物，只有少數地方有較多之植物種類；它生長之地方多屬於PH值3.0~3.5的酸性腐植土，土質粗鬆且土壤發育不良，有盤根的現象，土壤枯枝落葉層留有大量的台灣杜鵑枯枝落葉，人站在上面用力輕踩，會有晃動的感覺；杜鵑花屬共有十多種植物，台灣杜鵑是生長在台灣的固有種（endemic species）。

落葉樹與北部稜線的落葉林

■銅山附近台灣水青岡森林。　攝影／陳子英

國外進口的山毛櫸地板及家具是眾所喜愛的，事實上台灣也有稀有的水青岡純林，這種台灣水青岡也名列文化資產保存法中稀有保育類植物之一。廣泛分布於溫帶地區的水青岡屬（*Fagus*）植物在中更新世的冰期是台灣北部森林的主要優勢種，同時數量不只一種，冰河後退之後只留下台灣水青岡一部份留存在台灣北部的山頭上，另一部份的族群則留在大陸的華中，因此留在台灣的台灣水青岡也算是一種冰河的孑遺植物。

種類介紹

常綠闊葉樹

毽子櫟 *Cyclobalanopsis sessilifolia* (Bl.) Schottky

殼斗科 *Fagaceae*

毽子櫟屬於殼斗科的植物，葉邊緣為全緣，果實外面具有圓心輪的殼斗，植株能生長在風衝處，在台北縣、宜蘭縣衝風的稜線上形成5~7公尺高的低矮灌叢，尤其是在這一帶的稜線上，由於灌叢的枝椏密生，登山客一般很難通過；但在中海拔的森林中又形成喬木狀，甚至在新竹及桃園一帶的檜木林下又形成大喬木。

攝影／陳子英

栓皮櫟 *Quercus variabilis* Bl.

殼斗科 *Fagaceae*

栓皮櫟也是屬於殼斗科的植物，中海拔地區只出現於在中部及東部地區，"栓皮"是指植株的樹皮與台灣二葉松的樹皮相似，有厚且深縱裂的樹皮，但樹皮較台灣二葉松蒼白；由於栓皮櫟具有較厚的樹皮，同時根株下部能抽出新芽，在火災頻繁的地區，較能生存，因此也和台灣二葉松一樣都生長在較常有火災發生的地區。

攝影／陳子英

北部的落葉闊葉樹純林是指台灣水青岡在山頂或稜線所形成的夏綠林（summer green forest），分布在北部桃園縣的插天山及拉拉山一帶與宜蘭縣三星山以東的銅山、下銅山至鹿皮山一帶並形成純林，其他則零星分布於宜蘭縣阿玉山左側之山頭。

森林中伴生之闊葉樹有毽子櫟、水絲梨（*Sycopsis sinensis* Oliver）、新木薑子屬（*Neolitsea*）及灰木屬植物，林下地被層主要灌叢由玉山箭竹或瘤足蕨或紅苞鱗毛蕨（*Dryopteris subintegriloba* Serizawa）所構成。台灣水青岡雖然屬於純林，但下層的幼樹卻較少出現，宜蘭銅山地區和拉拉山地區直徑較大的母樹旁或玉山箭竹較稀疏的地區有幼苗出現，而阿玉山旁之山頭由於玉山箭竹較少，雖然植株數量較少，但不同徑級的台灣水青岡卻都有出現，由於台灣水青岡的幼苗較少天然更新值得進一步做監測。

■台灣赤楊。 攝影／陳子英

台灣赤楊又稱為"水柯仔"或"台灣榿木"，因此過去低海拔地區形成純林的地方，地名也有以柯仔林來命名的。台灣中海拔地區由於兩量豐沛，河川向源侵蝕，或由於人為開路，常形成面積大小不等的崩塌地或河川裸露地，這些崩塌地或裸露地由於地表裸露缺少土壤養分，因此只有早期能固氮的植物存活下來，在國外常有樺木（*Betula*）或白楊（*Populus*）等落葉樹的純林，在台灣只有台灣赤楊形成大片的落葉闊葉林，更特別的是赤楊的分布，在台灣中部多分布於中海拔山地，但在北部地區則降至低海拔闊葉林的河川裸露地或崩塌地。

中海拔山地除了台灣赤楊和台灣水青岡外，也有類似溫帶地區的其他闊葉樹，如山桐子、台灣胡桃、台灣檫樹、槭樹屬（*Acer*）、千金榆及櫻花屬（*Prunus*）等，但大多只點綴在中海拔森林間並不形成純林。

■台灣檫樹的樹皮縱深裂，形狀像二葉松的樹皮。

■台灣檫樹的幼樹分布在中北部的向陽山谷崩塌地或闊葉林樹倒所形的孔隙之中，葉是國蝶－"寬尾鳳蝶"的食物，尤其在棲蘭山及太平山一帶常混生於闊葉林中。 攝影／陳子英

長葉木薑子（南投黃肉楠） *Litsea acuminata* (Bl.) Kurata

樟科 *Lauraceae*

長葉木薑子屬於樟科植物，廣泛分布於台灣的中海拔森林，它不是冠層的突出樹（emergy tree），但在森林中的株數比例極高，尤其是林床下的幼苗非常多，與大葉楠的更新非常相同，在生殖更新上採取"人海戰術"，這與殼斗科植物略為不同，如長尾柯在林床下小苗較少，似乎採取"精兵政策"。

攝影／陳建志

山胡椒 *Litea cubeba* (Lour.) Persoon

樟科 *Lauraceae*

分布於全台灣中低海拔山區之闊葉林中，尤其多出現在開路兩側的崩塌地、森林伐採跡地、開墾地及闊葉林幾株喬木樹倒所形成的孔隙地。

屬於落葉灌木，葉膜質。全株具有刺鼻的薑辣香味，漿果成熟時為黑色，同樣具有濃厚的薑辣味，但原住民都在果實快接近成熟時採收它做為食鹽的代用品，新竹泰雅族的原住民稱之為"馬告（magow）"，為原住民的重要生活用品。

攝影／陳子英

人工柳杉林及檜木林

■明池附近的柳杉林。 攝影／陳子英

中海拔地區在光復初期至70年代為了木材的需求建造了許多柳杉林，柳杉（*Cryptomeria japonica* (L.f.) D.Don）原產於日本，在日本樹幹通直，木材為蓋房子做傢俱的上等材料，在台灣地區由於氣候比日本更溫暖，雖然年輪生長初期的速度比日本生長的快，但材質較鬆，因此無法像日本一樣當上等材料來用，柳杉林由於建造時是採取密植，加上隔一段時間即實施撫育，因此在林相的層次上通常只有上層的柳杉及林下的草本層，草本層中多以蕨類植物為主，主要有雙蓋蕨類、稀子蕨（*Monachosorum henryi* Chist）、瘤足蕨、鳳丫蕨類及蕁麻科（*Urticaceae*）的植物，如長梗紫麻（*Villebrunea pedunwlata* Shirai）、冷水麻（*pilea*）及赤車使者等。目前由於台灣地區遊憩的需求，許多遊樂區如溪頭及太平山及明池等都位於中海拔地區，柳杉林林相整齊許多林間步道都經過柳杉林中，因此柳杉林反而形成另一種景緻林。

在中海拔部份地區也將台灣原生的紅檜及台灣杉進行人工栽植造林，目前在許多地區也都漸漸成林，台灣杉的小枝條下垂，形態很像聖誕樹，但是由於葉子較尖銳，因此較無人採取做聖誕樹來裝飾。這二種樹木的材質在台灣地區都比柳杉好，但初期的直徑生長都比柳杉緩慢。

■福山附近的柳杉林。攝影／陳子英

墨點櫻桃 *Prunus phaeosticta* (Hance) Maxim. var. *phaeosticta*

薔薇科 Rosaceae

又名黑星櫻，與櫻桃、梅花及桃花等常見的果樹是同一屬的植物，但是果實不可食，葉子背面具有黑色腺點，葉子揉之，具有杏仁味的香味，是中海拔闊葉森林中極為普遍的植物，由於它的枝條在樹幹較低的地方也不脫落，因此樹形形成優美的圓錐形，目前已有開發做為園藝植物。

攝影／陳子英

台灣樹參*Dendropanax dentigerus* (Harms.) Merr.

五加科 Araliaceae

普遍分布在中海拔闊葉森林的中層，屬於五加科的植物，它的葉柄基部極為特殊，會膨大並把莖包一半，而且葉子有明顯的三出脈，所謂三出脈是指中肋兩側所斜側出的脈較明顯，加上中肋看似三條脈一齊由葉基部延伸出去；葉子的長法也很特別，有全緣不分裂的葉子，或裂成2或3裂深，和台灣木賨樹的葉子長法很相似，由於葉形的變異較大，在野外有時較難分辨。

攝影／曾彥學

烏心石 *Michelia compressa* (Maxim.) Sargent

木蘭科 Magnoliaceae

分布在中、低海拔的闊葉林中坡，在森林中幼樹的樹形高挑；但在全光照下，樹形形成圓錐狀，它與在車站或路邊買的玉蘭花都屬於木蘭科的植物，木蘭科是被子植物最出現的植物群之一，甚至恐龍存在的白堊紀就已經出現，木材早期就被使用做為廚房中的砧板，目前也有人將其木材做成精緻的傢俱，由於它在全光照下的圓錐形樹形極為優美，因此現在許多行道樹或遊樂區都有大量栽植。

攝影／曾彥學

落葉喬木

台灣胡桃 *Juglans cathayensis* Dode

胡桃科 Juglandaceae

歐洲的胡桃種仁是可以吃的，甚至部份的果實也可做成項鍊裝飾品；在戲劇上有以打開硬殼的"胡桃鉗"為名之戲曲。台灣地區在中部地區中海拔的河川地裸露地或開闊且潮濕的凹谷也有一種與國外類似的胡桃，名叫台灣胡桃，本植物之種仁也可以吃，但由於種子的分隔較多，種仁的肉較少較無食用的價值，無法像歐洲的胡桃一樣，大量栽培到市場販售，由於中海拔地區開闊且潮濕的凹谷或河川裸露地的生育地較少，同時許多地區都被開發來栽種果樹或農作物，因此台灣胡桃也成了稀有植物。

攝影／曾彥學

台灣檫樹 *Sasafras randaiense* (Hayata) Rehder

樟科 *Lauraceae*

幼樹分布在中北部的向陽山谷崩塌地或闊葉林樹倒所形的孔隙之中，葉是國蝶一"寬尾鳳蝶"的食物，尤其在棲蘭山及太平山一帶常混生於闊葉林中。植物體成長後形成又高又直的喬木，它的樹皮縱深裂，形狀像二葉松的樹皮，因此很容易與其他植物分辨，葉子的生長方式很特殊，葉邊緣形成全緣或2至3深裂，冬天會落葉，冬末春初開花的時候，也和梅花很像是先開花後長葉；花呈黃色，開花時整棵植株遠看像黃色的臘燭臺。由於木檫樹全株都有芳香味，而且木材屬於紅褐色，材質輕軟，因此也是傢俱的好材料，目前在太平山及棲蘭山一帶已有人工造林。

攝影／陳子英

青楓 *Acer serrulatum* Hayata

槭樹科 Aceraceae

攝影／曾彥學

分布在台灣的中、低海拔闊葉森林，幼樹時樹皮呈綠色。屬於槭樹科的植物，這一科的植物很多在冬天之際會落葉，同時在落葉之前會轉成紅色，但青楓只會形成淡黃色；它的葉子裂成五裂，而台灣低海拔地區之楓香（*Liquidambar formosana* Hance）葉子中裂成三裂，因此坊間常以「三楓五槭」來區分，事實上這種區分方式，並不正確，因為楓香屬國外也有葉子呈五裂或七裂的植物，而槭樹屬中的台灣三角楓（*Acer buergerianum* Miq.var. *formosanum* (Hayata) Sasaki），就屬於三裂的植物，兩屬的主要區別可以由葉序的生長方式及果實來區別，葉序的生長方式槭樹屬是對生，楓香屬葉子是互生，在果實上槭樹科為帶有薄膜翅膀的翅果，楓香的果實聚生成球狀的蒴果。

山桐子 *Idesia polycarpa* Maxim.

大風子科 Flacourtiaceae

分布在中海拔闊葉林的林線或河川的崩塌地，到冬天也會落葉。最容易讓人注意的是它具有長柄的心形葉，葉子的背面形成粉白色，特別是果實成熟時形成橙色紅之球形漿果並成串的掛在樹上，因此常常吸引各種鳥類來吃食，同時也是很好的觀果植物；木材早期是製造箱櫃及器具的材料。

攝影／陳子英

根節蘭屬植物

阿里山根節蘭 *Calanthe arisanensis* Hayata

蘭科 Orchidaceae

根節蘭屬（*Calanthe*）的植物是台灣中海拔闊葉林地被層很常見的植物，由於植物體的假球莖（pseudobulb）形成角錐狀，同時在角錐狀的假球莖上，有幾個節，因此稱為根節蘭，阿里山根節蘭是根節蘭屬中唯一分布在台灣的固有種（endemic species），阿里山根節蘭的形態優美，花徑中等，因此具有觀賞的價值，尤其又只生長在台灣地區，雖然分布全台灣而且族群數量也不少，但目前在野外也感受到人為採集的壓力。

攝影／陳子英

反捲根節蘭 *Calanthe puberula* Lindl.

蘭科 Orchidaceae

也是根節蘭屬的植物，是台灣中海拔北部闊葉林地被層很常見的植物，特殊的是它的花萼向下與柱頭形成90°的垂直角度，這在根節蘭屬中是極為特別的現象，與其他的根節蘭有很大的差異。

攝影／陳建志

其他草本植物

八角蓮 *Dysosma pleiantha* (Hance) Woodson

小檗科 Berberidaceae

八角蓮葉子邊緣裂成數個深裂，因此葉形成數個角狀，因而得名，不是所有的葉子都是八個角，但都是在八數的附近，植物體的根莖每隔一年長一節，當植株長到一定的大小，地上部才會有二片葉子，紅色的花形狀似鈴鐺狀，懸掛在二個葉柄的中間。八角蓮是有名的毒蛇解藥，再加上形狀特殊，在靠近都會附近的八角蓮多被大量採摘，在中海拔人跡罕至的闊葉林下，才可看到大量八角蓮的蹤跡。

攝影／陳子英

藤本植物

圓葉鑽地風 *Schizophragma fauriei* Hayata *ex* Matsum

八仙花科 Hydrangeaceae

中海拔在闊葉林的林緣或林冠孔隙中常可發現纏繞樹林的藤本植物有黃白色成叢的小花，事實這並不是真正的花，虎耳草科中鑽地風、八仙花屬的花序上都有特化成花瓣狀的花萼，真正的花朵非常細小，邊緣都圍繞有不同顏色與形狀的瓣狀萼片，這樣的生長方式較容易吸引昆蟲幫它授粉，在中海拔的森林邊緣，還有其他常見的藤本植物，如大枝掛繡球等藤本也有如此的特性，圓葉鑽地風的葉子為近闊卵形或卵狀圓形，因此有”圓葉”之名，它卵狀圓形的葉子可以與大枝掛繡球區分開來。

攝影／陳建志

長果藤*Aeschynanthus acuminatus* Wall. *ex* A.DC.

苦苣苔科 Gesneriaceae

在中海拔的潮濕闊葉樹林的樹幹上，常常可以看見樹上掛有一些藤本植物，長果藤就是其中很常見的植物，在樹上最容易讓人家印象深刻的就是它長長的莢果，他的莢果可長達15~20公分，開裂之後種子散開來，果莢仍會存留在枝條上一段時間，而它的花也很特別，很像進口的口紅花，花冠3裂片向後捲，雄蕊突出在花瓣的外面，雖然沒有口紅花那麼鮮艷的顏色，但黃褐色的花在背景是綠色的葉子前顯得格外突兀。

攝影／陳子英

神木輓歌

檜木林

The False Cypress Forest

◎撰文／楊國禎
◎ Text ／ Kuoh-Cheng Yang

中海拔地區是台灣雨量最多、濕度最高的地區，這個範圍經常形成雲霧，而在此雲霧裊繞的山區，正是檜木最適合的生長環境。檜木林的冠層喬木是針葉的紅檜或台灣扁柏，下層則是闊葉喬木、灌木及草本，也孕育了多種蕨類植物，林內的樹幹上密佈著生植物和蘚苔。在此一林帶中還有其他的針葉樹，如台灣黃杉、台灣杉、巒大杉、台灣粗榧、紅豆杉等，闊葉樹則主要是樟科和殼斗科植物。然而因爲檜木之材質優良，日治時代開始大量砍伐，如太平山林場、八仙山林場、阿里山林場等，即是爲砍伐檜木而存在的。直至目前殘存的檜木林已經不多，但卻仍有許多覬覦它的勢力存在著。

The mid-altitudes are the area of the most rainfall and the highest moisture in Taiwan. This area often sees fog, and where these fogs encircle the mountains are the environment most favored by the false cypresses trees. The canopy of false cypress forests is composed of Formosa cypress and Taiwan hinoki cypress, and the lower layers of broadleaf trees and shrubs and grasses. There are many species of fern as well, and the trees of the forest are tightly covered with moss. In this forest, there are also some conifer(gymnosperms), such as *Taiwania*, Taiwan Douglas fir, Ranta fir, Taiwan plum yew, and Chinese yew. The broadleaf species are primarily Lauraceae and Fagaceae. However, because of the high quality of timber, these forests were largely cut down starting during the period of Japanese rule, and the Taipin shan Forest Camp, Pahsien shan Forest Camp and Alishan Forest Camp existed solely to harvest these trees. By now, only a fraction of these forests remains.

■無意中留下的小雪山山巨木，可供人們憑弔原來檜木林的盛況。

攝影／楊國禎

■紅檜高大而長壽，多生長於山坡的下半段近溪谷的區域。 攝影／蕨類研究室

霧林

雲霧是潮濕空氣遇冷，空氣中多餘的水氣凝結成小水滴形成。高山島的台灣，承接東亞季風區恆常氣流，夏天由太平洋與南海吹向蒙古高原與西伯利亞，台灣首當其衝，冬天則相反的由蒙古高原與西伯利亞吹向太平洋與南海，吹到台灣前經過東海吸收水氣。當來自海洋潮濕的空氣遇山被迫上升後，水氣凝結成雲霧，這些雲霧在山區移動、翻滾，在海拔約 1,500~2,500 公尺間形成一恆常性的雲霧帶，北部地區的冬季會降低到不及 1,000 公尺。台灣中海拔的雲霧帶，雨量最多、雨日最長，濕度大、夏季涼爽、冬季降霜並偶而有雪。這種情況下，針葉樹與闊葉樹都可以生長，只有針葉樹而無闊葉樹生長的森林即為針葉樹林，只有闊葉樹而無針葉樹生長的森林即為闊葉樹林，如果針、闊葉樹混合生長則為針闊葉混淆林。森林外因雲霧帶來的水氣足以讓地衣類的松蘿生長，因而森林外或森林邊緣的樹上經常掛滿生長繁盛的松蘿；森林內則因溼度高，樹上經常掛滿苔鮮，而有苔林的稱號；林中巨大藤本植物不少，如大枝掛繡球（*Hydrangea integrifolia* Hayata ex Matsum. & Hayata）、愛玉子（*Ficus pumila* L.var. *awkeotsang* (Makino) Corner）、台灣長春藤（*Hedera rhombea* (Miq.) Bean var. *formosana* (Nakai) Li）；著生植物眾多，如小膜蓋蕨（*Araiostegia perdurans* (Christ) Copel.）、水龍骨科（Polypodiaceae）等蕨類植物；地被蕨類植物豐富，最具指標性的以台灣瘤足蕨（*Plagiogyria formosana* Nakai）為最，常見有複葉耳蕨類（*Arachniodes*）、瘤足蕨類（*Plagiogyria*）、雙蓋蕨類（*Diplazium*）、稀子蕨（*Monachosorum henryi* Christ.）、柄囊蕨（*Peranema cyatheoides* Hayata）、魚鱗蕨（*Acrophorus stipellatus* (Wall.) Moore）等，近溪溝 40 x 40 平方公尺的區域內常可達 50 種。

物種介紹

紅檜 *Chamaecyparis formosensis* Matsum.

柏科 Cupressaceae

台灣中海拔霧林的主要樹種，高大而長壽，以阿里山巨木為例，高 57 公尺、樹幹圓週 18 公尺、樹齡約 3000 年。屬於裸子植物的柏科，扁柏屬，是現存環北太平洋亞洲東部的台灣、日本，北美的阿拉斯加到美國西北及太平洋東岸的7個分類群之一，台灣特有種。種子萌芽後，小苗先有一段針狀葉的時期，以後則只長出鱗片狀葉。鱗片狀葉於莖上排列成十字對生的方式，先端尖銳而扎手，長約 0.2~0.3 公分，冬季經霜雪凍後呈現紅褐色。老樹常有很多分枝，頂端枝條常枯死而禿頂，樹幹基部常向下坡面延生出木板狀構造，用來支撐植物體的重量，分枝常生於背坡側，向坡側較少；樹皮紅褐色，號稱「薄皮仔」（相對於台灣扁柏的「厚殼仔」，閩南語），經常剝落，在樹的周圍經長期累積，堆積出厚而鬆軟的腐植層。毬果橢圓形，長約0.9公分，果鱗兩兩相對而生，每片果鱗有種子3~4粒，一棵生長良好直徑1.5公尺的獨立樹，一年可結約200萬粒種子，種子細小而輕，周圍有環形的薄翅，容易被風吹走。經常生長於山坡的下半段接近溪溝的區域。因生命期長，大樹生長位置的上緣常因樹木阻擋，大都侵蝕成緩坡，下緣無阻擋則流失成陡坡，紅檜巨樹林內的地表因而常呈現階梯狀。是百年來主要伐木對象之一，巨木常因蓮根菌腐蝕而樹幹中空，各林區都有少數巨樹因「無用之用」而留存供憑弔。

毬果橢圓的紅檜。 攝影／楊國禎

針闊葉混淆林

生長此帶的針葉樹有台灣鐵杉（*Tsuga chinensis* (Franch.) Pritz. var. *formosana* (Hayata) Li & Keng）、台灣雲杉（*Picea morrisonicola* Hayata）、台灣二葉松（*Pinus taiwanensis* Hayata）、紅檜（*Chamaecyparis formosensis* Matsum.）、台灣扁柏（*Ch. obtusa* Sieb. & Zucc. var. *formosana* (Hayata) Rehder）、台灣黃杉（*Pseudotsuga wilsoniana* Hayata）、台灣肖楠（*Calocedrus formosana* (Florin) Florin）、台灣華山松（*Pinus armandii* Franchet var. *masteriana* Hayata）、台灣五葉松（*Pinus morrisonicola* Hayata）、台灣杉（*Taiwania cryptomerioides* Hayata）、巒大杉（*Cunninghamia konishii* Hayata）、台灣粗榧（*Cephalotaxus wilsoniana* Hayata）、紅豆杉（*Taxus sumatrana* (Miq.) de Laub.）等。其中台灣鐵杉、台灣二葉松常形成純林或與其他樹種混生成針葉樹混生林、針闊葉樹混淆林，紅檜、台灣扁柏、台灣雲杉偶而形成純林，但大部分形成針葉樹混生林、針闊葉樹混淆林，台灣黃杉、台灣肖楠則於曾崩塌的陡峭岩壁形成小面積純林或針闊葉樹混淆林，台灣華山松、台灣五葉松、台灣杉、巒大杉是組成針葉樹混生林、針闊葉樹混淆林的主要大樹的種類，紅豆杉以大樹散生於闊葉林中，台灣粗榧散生於闊葉林下層。

■香杉大樹的樹姿。　攝影／楊國禎

個體數量最多、分布最廣、材積最大為同兄弟的紅檜和台灣扁柏，故此帶常稱為檜木林。

針葉樹高約20~60公尺，闊葉樹則約20~35公尺，形成針闊葉樹混淆林時，針葉樹佔第一層，闊

台灣扁柏 *Chamaecyparis obtusa* Sieb. & Zucc. var. *formosana* (Hayata) Rehder

柏科 Cupressaceae

台灣過去伐木期間最受歡迎而貴重的用材，木材濃郁的精油香氣曾經是所謂「酚多精」的主產品；與紅檜並稱檜木，因香氣濃、樹幹通直、比較不會空心、生長在山稜頂部，受到的砍伐比紅檜更嚴重。因樹皮厚而號稱「厚殼仔」(閩南語音)，與樹皮較薄的紅檜(薄皮仔)相對稱呼。細小不及0.2公分的鱗片狀葉在莖上十字對生，而且將莖包裹在裡面，莖較圓、鱗片狀葉較鈍、枝條背面的葉間常有白粉、果正球形，這些特徵也與紅檜的莖較扁、鱗片狀葉較尖銳、枝條背面的葉間不具白粉、果橢圓形相對的不同。棲蘭山區是目前世界僅存面積最大的台灣扁柏林。

攝影／蕨類研究室

■球果正圓形的台灣扁柏。　攝影／楊國禎

台灣杉 *Taiwania cryptomerioides* Hayata

杉科 Taxodiaceae

首先在台灣發現，因而將Taiwan拉丁化成*Taiwania*當做屬名，種名*cryptomerioides*的意思是類似*Cryptomeria*（柳杉，台灣中海拔伐木後大量造林）的植物，是當時重要而盛大的事，紀錄最高的有90公尺，是台灣最高大的樹，60~70公尺高度的樹非常常見。樹幹通直，樹形優美，約20公尺以下幼樹的小枝條向下垂、葉子是尖銳的針刺形，長為大樹後的小枝條變成往上舉、葉子變成鱗片狀，是會變身的植物。台灣杉屬的植物在地質年代時普遍生存於北半球，是種類非常繁盛的類群，現孑遺於東亞的台灣至雲南間的山區，呈現零星分布的狀態，在台灣則零散分布在檜木林中，在高大的檜木林中更顯的突出。2002年2月證實在台灣南部雙鬼湖山區有大片、密集而壯觀的森林。

攝影／蕨類研究室

■結毬果的台灣杉老樹枝條，葉呈鱗片狀。　攝影／楊國禎

葉樹大喬木則佔第二層、第三層是小喬木、第四層是灌木、第五層為地表面草本。

針葉樹林主要生長北溫帶地區，在台灣分布約由海拔3,600公尺到海拔1,500公尺左右， 2,500公尺以上是針葉林的區域，且由單一樹種形成純林，約3,600~3,100公尺是台灣冷杉林，約3,100~2,500公尺是台灣鐵杉林，1,500公尺以下的針葉樹林是很特殊且少有的。闊葉樹林由低海拔往較高海拔分布，2,500公尺幾乎是闊葉樹林分布的上界。海拔1,500~2,500公尺間是針葉樹與闊葉樹都可以生長的範圍，形成針闊葉混淆林。

更新

針葉樹常是不耐遮陰的植物（目前構成森林的主要針葉樹種只有台灣扁柏較耐遮陰），必須於崩塌地、火災跡地或枯倒樹形成的森林破空（洞）等等陽光較強區域，小樹苗才能由種子發芽而建立、生長，且長成森林，但很多闊葉樹是耐遮陰植物，在森林底下小樹苗可以萌芽建立，並逐漸長成大樹，或等森林破空陽光大量灑入時，小樹苗迅速長大成大樹。因此針闊葉混淆林中除了第一層大樹外，森林下常不見針葉樹小樹或小苗，當環境不再有崩塌、火災等除去森林的變動發生，闊葉樹會逐漸侵入針葉樹林中，等針葉樹逐漸死亡，森林也逐漸轉變成闊葉樹林。耐遮陰的常綠闊葉樹種類多且生長旺盛，並能在森林下建立小苗更新，壽命雖只有數百年，但能一直代代更替而永存，而針葉樹雖然是陽性樹種，但生命期往往長達千年以上，森林下常綠闊葉樹常已侵入而形成針闊葉混淆林。因此，生長快速而壽命短【如台灣赤楊（*Alnus formosana* (Burk.) Makino）常不及百年】的陽性落葉闊葉樹生長的空間被壓縮為崩塌地早期，因而在沒有人類活動的原始林時代並不發達，但隨著人類開發的頻繁，落葉闊葉樹數量增多，不僅森林齡級年輕化，更是森林品質劣化的象徵。

台灣黃杉 *Pseudotsuga wilsoniana* Hayata

松科 Pinaceae

零星分布於中海拔河岸的高大針葉樹，往往以高40~50公尺的大樹呈現，小苗與幼樹生長崩塌而陽光強的岩壁陽坡，生育地與檜木區隔開，現在主要分布於大漢溪上游的鎮西堡地區、大甲溪上游的梨山地區、高雄六龜的出雲山區、臺東南橫的霧鹿至美奈田主山間。因葉形「長條片狀」而被冠上「杉」名，但毬果是松科植物標準的特徵：苞鱗與果鱗分開。特殊的是，像「戟」三叉的苞鱗片片露出於毬果外，好像蛇的舌頭吐「迅」但收不回去的樣子，模樣非常奇特；而長約10公分左右的長橢圓形（或稱長紡槌形）毬果，成熟後由綠轉褐而後變為黑色，一顆顆垂掛在枝條上，加上通直而有縱裂條紋的樹幹，構成台灣黃杉的特徵。

巒大杉（香杉）*Cunninghamia konishii* Hayata

杉科 Taxodiacee

零星散布於檜木林中，是東亞孑遺的類型之一，葉是條片狀而向先端漸尖的披針形，形狀稍微彎曲而一直以常見用來割草的鐮刀形狀來比擬，這是杉科植物的特徵，與松科植物葉條片狀，先端不漸尖(松樹類葉針狀)的主要差別之一，另一差別是杉科毬果的果鱗與苞鱗合生而分不開，松科毬果的果鱗與苞鱗分開如台灣黃杉。原生於台灣的族群與產自中國的族群在葉形上有大（原產自中國的杉木）、小（巒大杉）的差別，百萬年來冰河期來來去去造成的基因交流、各自演化，這是不易切割開的差別，加上研究者生長背景不同所影響個人價值觀的差異，要釐清通過事實的存在與人為的判斷間，這糾葛不清的縱錯複雜關係，下一個單一價值判斷的結論，在這價值多元的年代實在不是件容易達成的事。

暖溫帶及冰河期孑遺植物的避難所

針葉樹開始出現於石炭紀，侏儸紀、白堊紀等恐龍繁盛時代的綠色世界，主要由針葉樹構成，但當恐龍滅絕後的第三紀以降，闊葉樹林由熱帶地區興起，並逐漸向溫帶地區侵入，很多針葉樹的生長區被闊葉樹佔去了，繁盛的種類逐漸消失、滅絕，成為全球性相對稀少的類群。再加上地質史最後的兩百萬年來（第四紀），發生全球多次大規模變冷、變熱的交替，變冷時期（冰河期）北半球植物必須南遷，北方針葉樹南遷幅度更大，形成三明治式的壓縮效應，針闊葉混淆林帶的針、闊葉植物同遭壓縮，大量的個體和種類因而消失，變熱時期(間冰期)闊葉樹向北擴張，針闊葉混淆林帶的植物被推著往北移。在此南來北往的遷移中，很多種類適時加入，也有物種退出，演變成複雜的綜合體。

位於東亞陸塊東南邊緣的台灣，不僅氣候適宜、位置適於保留路過者、4,000公尺縱深的垂直變異緩衝抵得過緯度40度的變化，而且山區多變的微地形氣候差異，以及高度變動的地體，創造出更多具時、空差異的生態棲位，保留住這批落難者。台灣的中海拔就成為暖溫帶及冰河期孑遺植物的主要避難所之一，也是台灣生物多樣性基本樣貌展現的一部份。舉幾個例子：前面提到此帶保存13種台灣分布廣泛與數量較顯著的針葉樹，如加上生態幅度較廣泛的刺柏（*Juniperus formosana* Hayata），邊緣侷限分布的馬尾松（*P. massoniana* Lamb.）、清水圓柏（*J. chinensis* L. var. *tsukusiensis* Masamune）、台灣穗花杉（*Amentotaxus formosana* Li）、台灣油杉（*Keteleeria davidiana* (Franchet) Beissner var. *formosana* Hayata），合計有 18 種，已遠超過台灣裸子植物種類的一半以上（18/28）；而鐵杉、扁柏、黃杉、肖楠、台灣杉、杉木、粗榧、紅豆杉、穗花杉、油杉這些屬（Genus）都是著名的孑遺而侷限分布的裸子植物類群；只侷限分布台灣、日本的單種科植物一

紅豆杉 *Taxus sumatrana* (Miq.) de Laub.

紅豆杉科 Taxaceae

因體內存有能治癌症的化學成分—紫杉醇(taxol)，因而聲名大噪，從原本不值一顧的情狀，身價爆漲。是獨立成科的紅豆杉科植物，世界上孑遺存活下來的種類不多，廣泛而零星的散生於台灣中海拔的森林中。種子成熟時外面包著一層紅色稱為「假種皮」的肉質構造，這構造長自種子基部的柄上，因而名字有「紅豆」的字眼，也因葉子是條片狀而被冠上「杉」名，成為樹上長有紅豆而葉子條片狀的溫帶植物。但既不結毬果而種子外露，葉背的氣孔帶不顯著也非白色，枝條對生更非螺旋狀的互生排列方式，整體來說與杉科植物差別甚大。

森氏櫟 *Cyclobalanopsis morii* (Hayata) Schott.

殼斗科 Fagaceae

■葉柄長的森氏櫟，堅果正在成長中。

因文字與習慣的流傳，溫帶的殼斗科植物傳統上以歐美常見的落葉樹廣為大眾所熟知，習俗上稱這一群為橡樹或櫟樹（oak），是冬天落葉的大喬木，葉子大型具有芒刺狀的鋸齒或開裂的邊緣，開花時樹上掛滿下垂的雄花序，它會放出滿天飛舞的黃色花粉，結的是松鼠愛吃而圓胖的硬果實（稱堅果Nut），這個果實被一個一小片、一小片拼湊成的盤狀構造托著，木材是就地取材的上好建築材料。但在濕潤而獨特的東亞地區，發展出一群由這些特徵而稍作轉變的類群；春天長新芽時才把去年的老葉落光光，感覺上是一年到頭都有葉子的常綠樹；葉子細小而堅硬，邊緣不是沒有鋸齒就是只有一半而且不會開裂，托著果實的盤子是由一圈圈圓盤所組成的。這群生長在台灣中海拔的變形橡樹(櫟樹)，以森氏櫟的數量最多，最具代表性，長長的葉柄撐著基部接近圓形的葉子，葉子的邊緣由基部的滑順（全緣）逐漸變寬，到達最寬部位（約為葉子長度的2/3)時轉變為尖銳的鋸齒緣，並突然縮收到先端成尖尾巴，葉子的背面是光亮的綠色，乾後也與正面一樣變成咖啡色，是針闊葉混合林中構成 20~30 公尺高闊葉樹層次的最主要樹種。

攝影／蕨類研究室

昆欄樹（*Trochodendron aralioides* Sieb. & Zucc.）是此帶的代表優勢樹種之一；五茄科（Araliaceae）與四照花科（Cornaceae）植物大多呈現寡種屬的孑遺狀態，分布台灣此帶的有食用土當歸(*Aralia cordata* Thunb.）、台灣樹參（*Dendropanax pellcidopunctata* (Hayata) Kanehira ex Kanehira & Hatusima）、台灣八角金盤（*Fatsia polycarpa* Hayata）、台灣常春藤（*Hedera rhombea* (Miq.) Bean var. *formosana* (Nakai) Li）、台灣五葉參（*Pentapanax castanopsisicola* Hayata）、台灣鵝掌柴（*Schefflera taiwaniana* (Nakai) Kanehira）、華參（*Sinopanax formosana* (Hayata) Li）、通草（*Tetrapanax papyriferus* (Hook.) K. Koch）、東瀛珊瑚（*Aucuba japonica* Thunb.）、四照花（*Benthamidia japonica* (Sieb. & Zucc.) Hara var. *chinensis* (Osborn) Hara）、台灣青莢葉（*Helwingia japonica* (Thunb.) Dietr. ssp. *formosana* (Kanehira & Sasaki) Hara & Kurosawa）、燈臺樹（*Swida controversa* (Hemsl.) Sojak）、梾木（*Swida macrophylla* (Wall.) Sojak）等計有13種，幾全部可列為世界性稀少的類群。

森林之最

最長壽：紅檜，約3000年（阿里山巨木，已倒），台灣扁柏也差不多長壽，一般可達2000年以上。

最大：紅檜，大部分沿山坡生長板狀構造，實測大都不準，現知最大實際圓週約20公尺（司馬庫斯巨木），直徑近7公尺。

最高：台灣杉，紀錄90公尺，林業人員相傳超過100公尺並不稀奇，常見達70公尺以上，如八通關古道對關附近、棲蘭林道160線底、大鬼湖北本野山前。

檜木林：平均高度可達40~50公尺，樹幹直徑大都2公尺以上，是台灣最高大森林，現存較大片者有桃園拉拉山巨木區、新竹鎮西堡巨木區、棲蘭林道170線、南橫檜谷區等。

昆欄樹 *Trochodendron aralioides* Sieb. & Zucc.

昆欄樹科 Trochodendraceae

昆欄樹是打破傳統教育上一貫「二分說法」的好例子。顯花植物(開花植物，有花植物)類群皆有導管、花萼花瓣，這是教科書上常見的說法，但屬於顯花植物的昆欄樹沒有導管、花萼花瓣，就必須把他當例外來處理。本來所有事項就沒有一定都要齊頭並進，有的快、有的慢是很自然的事，聞道有先後，術業有專攻嗎！脊椎動物的四肢，在魚類未出現，鳥類前肢是翅膀，鯨呈鰭狀，蛇類稱為退化，植物也有類似的情形！特殊的是，這一科就只此一種單傳下來，只生存於太平洋西岸的台灣到日本的臨海島鏈上，也是台灣霧林的指標樹種。油亮光滑的外表，枝葉層層相疊組成層狀結構的樹形，有著長葉柄的葉子在條頂端簇生成叢，葉柄長度由整叢最長的外圍向內逐漸縮短，雌蕊排成圓盤狀。個體依雌雄蕊成熟的時間可以分成兩型，雌蕊先成熟個體的雄蕊約晚一個月成熟（稱雌先熟植株），同時有些個體的雄蕊先成熟，相對的雌蕊約晚一個月成熟（稱雄先熟植株），可以把花粉傳給前面說的「雌先熟」植株，「雌先熟」植株的晚成熟雄蕊的成熟花粉就傳給「雄先熟」植株晚成熟的雌蕊，在「雌雄同株」的情形下，形成功能上如同「雌雄異株」的種類，同樣必須異株受粉。

華參 *Sinopanax formosana* (Hayata) Li

五加科 Araliaceae

目前尚被承認的台灣少數特有屬之一，是五加科東亞孑遺類型的最侷限分布型，但在台灣的分布卻是全面而零星的廣泛散佈於中海拔山區，是林下的大灌木至小喬木，大形葉的邊緣成多角狀的大鋸齒，葉背被棉厚而灰白色至褐色的銹毛，花序的第一級、第二級分枝成總狀花序型，具有五加科植物的基本特色：分枝甚粗壯，葉柄基部形成鞘狀把莖包住，掌狀形的大型葉，花序的最末一級分枝排列成繖形，花小。

百年伐木與現今保育行動

檜木林近百年來因材質優良、蓄積量大，成為瘋狂砍伐的對象，由日據時代的太平山、八仙山、阿里山所謂的三大林場的砍伐，至國民政府時代逐漸擴張為幾乎全島性的殺戮戰場，伐木的林道密密麻麻蜿　穿梭於中海拔山區，至民國70年代則幾已砍完，民國78年林務局以伐木來自籌經費的預算已入不符出，全面改為公務預算，僅存的檜木林依賴著陡峭易崩塌的地形或藉著石門水庫上游集水區保安林的劃設而倖免於難。然而退輔會森林開發處（民國89年更名為森林保育處）在須自籌經費並需盈餘繳庫的壓力下，編造枯立倒木妨礙森林更新為由，繼續「處分」(林業用語，伐木就像處罰犯錯的人一樣)僅存於棲蘭100號林道西邊最大片的原始檜木林，這片檜木林是因民國64年葛樂禮颱風造成石門水庫上游大量崩塌、淤積嚴重，因此劃設為水庫上游集水區的保安林，如此才逃過被砍伐的命運而存留下來。

鑒於汐止多次的淹水，及1997年賀伯的土石流災變，問題源頭都指向百年來過度砍伐森林造成，民間保育團體因而發起搶救棲蘭檜木林運動，作為搶救山林的指標，連續三年發動為子孫、為未來而走的森林抗暴街頭遊行。然而行政院一面安撫保育團體，一面核准所謂棲蘭永續當林示範區，繼續「處分」這片石門水庫上游保安的原始檜木林。在2000年的象神颱風再度肆虐汐止，2001年桃芝再度重創南投山區，2002年上半年的嚴重乾旱後，期望在一片砍檳榔與全民造林聲中，全民還能清醒，百年的創傷與罪孽，需用千年來彌補，保障台灣千年安定的中海拔原始檜木林，才是台灣免於土石流與旱災威脅的保安符，「人類的作為」必須全面退出原始檜木林，以及檜木林生存的保安區域，讓台灣山區得以生養孳息。請為這段歷史留下註記。

鳳仙花 *Impatiens* spp.

鳳仙花科 Balsaminaceae

鳳仙花類是一群奇特的肉質小草本植物，種子靠果實裂開的彈力彈出、傳播，當碰觸成熟果實時，果實突然裂開將種子射出，這種狀況令人覺得驚異，所以英文稱之為「Touch-me-not」。花也很特殊，常形成囊狀並於尾端拉長成細管，這細管子特稱為「距」，因為囊膨大的關係使花的開口歪向一邊，配上囊與「距」，側看時形狀很像一隻昆蟲，花的開口處像昆蟲的頭部，囊與「距」就是腹部，整個花的構造正適合某種型態的昆蟲鑽進去吸花蜜，順便傳粉，花蜜就位於「距」的位置，「距」的長度與大小限制了昆蟲的形態，要有能配合的口器的昆蟲才能吸的到花蜜，不同的鳳仙花就有不同的昆蟲搭配。台灣原生的鳳仙花有三種，紫花鳳仙花（或稱單花吊船花）(*Imp. uniflora* Hayata）、黃花鳳仙花（*Imp. tayemonii* Hayata）、棲慕華鳳仙花（*Imp. devolii* Huang)，它們全部都是台灣特有種，也都生長於中海拔的檜木林帶，但是不同種分布的情況不一樣，相當奇特。紫花鳳仙花普遍生長於中海拔森林邊緣山溝或稍有光線的森林下濕潤地，屬於全面分布型；黃花鳳仙花只生長於台灣北部山區，且分布點相對稀少；棲慕華鳳仙花則只分布於新竹觀霧山區，且數量非常密集，甚為奇特，更特殊的是觀霧地區三種鳳仙花都有。會這樣分布的原因值得探討！

上■棲慕華鳳仙花
中■紫花鳳仙花(或稱單花吊船花)
下■黃花鳳仙花。

攝影／蕨類研究室

冰河時期的難民

昆欄樹

Bird-Lime Tree, *Trochodendron aralioides*

◎撰文／吳瑞娥
◎ Text ／ Juei-Er Wu

昆欄樹目前僅間斷分布於台灣、琉球群島及日本等地，惟不論就分布地點或數量而言，其主要族群在台灣而非日本或琉球群島。在台灣，昆欄樹廣泛分布於環中央山脈的中海拔霧林帶地區，與台灣扁柏、紅檜混生。昆欄樹爲東亞特有植物，這一類植物在第三紀古新世前期就出現於格陵蘭東部的地層中，被認爲是第三紀的孑遺植物，而且在過去的地質年代中曾有遠較現今爲廣的地理分布範圍。根據昆欄樹及其相關類群之歷史地理分布情形，昆欄樹可能起源於中國西南一帶，在台灣各地的族群中，以中部地區累積較多遺傳變異且具有較多古老的祖先型。推測該區域可能是昆欄樹於第四紀冰河時期的避難所，冰期後再由此散佈到其他各地。

Trochodendron aralioides Sieb. et Zucc. is currently found only in Taiwan, the Rhykyus and Japan. Based on its distribution and numbers, the primary concentrations of this plant are found in Taiwan, rather than in the Rhykyus or Japan. In Taiwan, *Trochodendron* is widely distributed in the mid-altitude forest belt of the central mountain range and lives together with Taiwan cypress and Taiwan red cypress. *Trochodendron* is endemic in Southeast Asia. They had already appeared by the Tertiary in the eastern part of Greenland. They are thought to be a relic species from this age, and in past geological ages must have been much more widely distributed than they are today. Based on historical analysis of the distribution of *Trochodendron* and related species, *Trochodendron* is thought to possibly have originated in the southwestern part of China. Of the populations in Taiwan, those in the central region have accumulated more genetic variation and show more ancient. It can be deduced that this area has long been a favored growing environment for *Trochodendron*, and may have provided a refugium for this tree during the Quaternary when glacier once again covered many regions.

■昆欄樹主要生育於霧林帶。

攝影／楊國禎

每次站在陽明山的昆欄樹純林下，內心總是悸動不已，常納悶的想著：這種溫帶樹種怎麼會長在這裡呢？

原始的植物

昆欄樹科(Trochodendraceae)僅含一屬一種，即昆欄樹（*Trochodendron aralioides* Sieb. & Zucc.），為東亞特有植物。昆欄樹為高大常綠闊葉樹種，葉菱形至披針形，具光澤，花序為頂生之總狀花序，雄蕊多數，排列成輪，心皮多數約5~10個，呈環狀排列，且心皮僅在基部相連，又木質部中缺乏導管，僅具假導管，故一般認為是被子植物中較為原始的種類。

■昆欄樹花的構造很簡單，中心之雌蕊排成一圈，雄蕊環生成數輪，其外無花被。 攝影／王震哲

根據化石紀錄，昆欄樹類植物早在第三紀古新世前期（Paleocene）就出現於格陵蘭東部的地層中，被命名為擬昆欄樹(*Tetracentronites hartzi*)；在始新世（Eocene）中晚期中國遼寧省撫順的煤系地層中，也曾找到一種昆欄樹(*Trochodendron* sp.)的葉化石。故此類植物被認為是第三紀的孑遺植物，而且在過去的地質年代中曾有遠較現今為廣的地理分布範圍。

植被壓縮分布現象與棲地異質性

昆欄樹目前僅間斷分布於台灣、琉球群島及日本等地，惟不論就分布地點或數量而言，其主要族群在台灣而非日本或琉球群島。在台灣，昆欄樹廣泛分布於環中央山脈的中海拔霧林帶地區(cloudy zone)，與台灣扁柏（*Chamaecyparis taiwanensis* Masamune & Suzuki）、紅檜

(*Chamaecyparis formosensis* Matsumura)混生，此區域屬於暖溫帶山地針葉林。其年雨量約3,000 ~ 4,300公釐左右，相對濕度全年平均80%，為本省雨量最多、大氣濕度極高之地區。在台灣中部其分布大都在海拔2,000公尺以上，在台灣南部的分布可低至海拔1,000公尺以下，而在台灣北部目前所知甚至可分布到接近海平面。明顯可見此溫帶樹種受植被壓縮的影響而與熱帶樹種共域分布的現象。

在日本，昆欄樹呈零星分布於北緯37°以南，海拔300~2,000公尺的山區，年雨量約1,250公釐以上區域。在琉球群島僅分布於近台灣的西表島(Iriomote Island)及近日本本土的庵美大島(Amami-O-shima)，而最大的沖繩島並無昆欄樹，呈現侷限分布現象。日本及西表島族群主要分布於溪谷地形，與台灣主要生育於山腹之情形顯然不同，分布於台灣與日本的族群間似有明顯的棲地異質性存在。但整體而言，從昆欄樹喜分布於霧林帶(台灣)及溪谷(日本)的特性，以及台灣族群呈現與雨量密切相關的現象來推論，濕度對昆欄樹應是一重要的選汰因子。

■昆欄樹花序。

昆欄樹之現況與保育措施

昆欄樹目前僅侷限分布於台灣、琉球及日本，而其相關類群，如水青樹（*Tetracentron*）、連香樹（*Cercidiphyllum*）、領春木（*Euptelea*），現今也只分布於東亞地區，且以中國西南至華中一帶為主要分布範圍。根據前述昆欄樹及其相關類群之歷史地理分布情形，推測昆欄樹可能起源於中國西南一帶，於第三紀時擴散至亞洲大陸東緣，後來因台灣在中新世（Miocene）時發生造山運動，中央山脈隆起，或許正提供昆欄樹於冰期交替期間的良好棲所。

攝影／王震哲

藉由等位異構酵素電泳法與DNA片段定序分析的結果，發現台灣與琉球南部西表島的昆欄樹族群間親緣關係較近，而與日本本島及琉球北部奄美大島族群間之親緣關係較遠，顯示在兩大族群（台灣與日本）之間由於地質年代上長久的隔離，造成明顯屬於地理割裂的隔離分化（Vicariance）。至於在台灣各地的族群中，以中部地區累積較多遺傳變異且具有較多古老的祖先型。推測該區域長久以來為昆欄樹族群較佳的生長環境，可能是昆欄樹於第四紀冰河時期的避難所，於冰期後由此再散佈到其他各地。昆欄樹雖為侷限分布於東亞島嶼之孑遺植物，但在台灣與日本兩大棲地的遺傳變異度皆高於其它的地理侷限種與特有種，顯示該種植物仍能藉基因流傳引入一些新的基因，故良好的棲地維持應為最佳的保育措施。

■昆欄樹之果實。

■昆欄樹之果實成熟時裂開，露出微小的線形種子。

攝影／王震哲

風姿綽約萬人迷

野生蘭

Taiwan's Wild Orchids

◎撰文／鄭育斌、呂勝由
◎Text／Yu-Pin Cheng
Sheng-You Lu

蘭科植物是維管束植物中最進化也是最美麗的一群植物，主要分布於熱帶和亞熱帶地區。台灣位居歐亞大陸東緣，包括熱、暖、溫、寒垂直氣候帶，加以複雜的地形變化，因而蘊含了大約340種的各類原生種蘭花。依照生長型可區分爲地生蘭、附生蘭、腐生蘭等。台灣原生的蘭科植物約佔台灣維管束植物的十分之一，爲台灣的第一大科植物，以單位面積而言，也遠較鄰近地區爲高。蘭科植物大都具有優雅的外型，特別是它的花朵經過高度的特化之後，造型各異，同時也十分迷人。因爲蘭科植物的觀賞性高，具有高經濟價値，因而也遭到空前的商業性濫採，造成族群數量的銳減。當務之急是加強蘭科植物的保育工作，使寶島繼續擁有鳥語花香的美譽。

Orchids are the most evolved and the most beautiful of the vascular plants, and are mainly distributed in the tropical and subtropical regions of the world. Taiwan is located at the eastern edge of the Eurasian continent, and includes tropical, warm, temperate and arctic climates. If we add the island's complex topography, we see why Taiwan contains approximately 340 of the world's wild orchid species. Orchids can be classified as ground, epiphyte, or saprophyte according to their habits. Taiwan's wild orchids account for some 10% of all vascular plant species in Taiwan. They are the largest family of plants in Taiwan, and in terms of unit area, the area of orchids in Taiwan far surpasses the next-ranked area. Orchids are mostly delicate in appearance, particularly in their flowers, which have undergone a high degree of adaptation. These are of unusual shapes, and are very alluring. Because orchids have a high ornamental value, their economic value is likewise high. Consequently, they have been over-harvested to a point never before seen. This has created a reduction in their numbers, and the pressing task before us today is to enhance our conservation efforts for the orchids. In this way we can ensure that our island will always have the lovely reputation of these fragrant flowers.

■台灣喜普鞋蘭特產於台灣的中高海拔山區。

攝影／呂勝由

蘭科植物（Orchidaceae）是維管束植物中最進化的一群，全世界約有24,000種之多，除了極區和沙漠之外，地球上的其他區域都有蘭科植物的分布。尤其是熱帶和亞熱帶地區，因為氣溫和濕度適宜，蘭科植物蓬勃生長種類特別多。而台灣正巧位於北回歸線通過的亞熱帶，高溫多濕的環境和綿延蓊鬱的森林，非常適合蘭花的生長，從海平面至最高峰玉山山巔，都有蘭科植物的分布。根據台灣植物誌的記錄，台灣的原生蘭科植物約有340種，種數幾近台灣維管束植物的十分之一，是台灣維管束植物組成的第一大科。

蘭科植物是多年生的草本植物，它的生態、型態和習性等，千變萬化，具高度的歧異性。根據生活型通常可將蘭花分為地生蘭、附生蘭或腐生蘭等。溫帶地區大部份為地生蘭，而熱帶和亞熱帶地區，則有很高的比例為附生蘭，尤其在熱帶雨林區，茂密的森林提供更多的棲位（niche），高大的樹幹上往往佈滿了各式各樣的蘭花。腐生蘭是不具葉綠素的蘭花，在比例上較少，這類蘭花無法自己行光合作用，靠吸收其他植物分解物生存，出現的時間非常短暫，通常只在某一季節就完成生活史，是相當特殊的蘭花。

蘭科植物之所以特殊，除了型態之外，最重要的是它的花高度特化性。蘭科植物的花大小不一，通常為左右對稱花，最吸引人的是地方是唇瓣，它也是分類上最重要特徵。另外蕊柱的發育也是蘭科植物的關鍵特徵，一般植物花部的雌雄蕊都是分離的，蘭科植物的雌雄蕊合生成為蕊柱，這種特殊的構造有助於授粉工作的完成。

台灣的蘭科植物研究或採集，最早開始於1878年英國人Arthur Corner第一次發表台灣的新種蘭花一鳳蘭（*Cymbidium leachianum* Reichb. f.）和黃絨蘭（*Eria corneri* Reichb. f.），隨後許多歐美學者，如英國的Henry等人也曾到台灣採集。日據時期的台灣，大量的蘭花新種被發表，最具代表性的人物是早田文藏（Hayata），幾乎一半種類的台灣野生蘭都是由他命名。其他的日本學者如山本由松（Yamamoto）和正宗嚴敬（Masamune），後期的福山伯明（Fukuyama）和瀨川孝吉（Segawa）等，都對台灣的野生蘭研究有非常大的貢獻。

位於亞熱帶的台灣，蘭花的組成非常多樣，野生蘭不僅在數量或種數上都非常的豐富，兼具各種生活類型的蘭花。和鄰近地區比較，台灣的蘭科植物的單位面積種數，遠高於其他鄰近地區。就植物地理而言，大武至浸水營一線以北地區組成和日本、琉球和中國類似；以南則和菲律賓或南洋類似。所以台灣正位於植物分布的交會帶，重要性不可忽視。近年來的森林破壞和人為的採集，對台灣的野生蘭造成非常大的傷害，不僅是具觀賞價值的野生蘭，連一些常見種類都逐漸在野外消失了。為呈現台灣野生蘭的多樣性，本文挑選許多常見或是具代表性的野生蘭介紹如下。

■台灣白及是低海拔或平地常見的野生蘭，常見於陽光充足的公路邊坡和草生地。 攝影／呂勝由

種類介紹

攝影／鄭育斌

綬草 *Spiranthes sinensis* (Pers.) Ames

蘭科 Orchidaceae

地生蘭，高約20~30公分，具有粗壯、簇生的肉質根。莖甚短，葉線形，肉質。花莖10~25公分長，穗狀花序，小花在花軸上螺旋排列，粉紅色，少數全白。綬草是台灣地區最早被紀錄的野生蘭花，早在1863年英國人Swinhoe在List of Plants of the Island of Formosa，就紀錄一種蘭科植物名為*Spiranthes australis* Lindl.，其實就是現在的綬草。它主要分布於低海拔平原草地或山坡地，在市區校園內的草地上也常常可以見到它的蹤跡，近年來因為人為採集供為藥用，導致族群日益縮小。

攝影／呂勝由

蝴蝶蘭 *Phalaenopsis aphrodite* Reichb. F.

蘭科 Orchidaceae

世界知名的附生蘭，被稱為「蘭花之后」，在翠綠肥厚的葉片襯托下，一串串白色的花朵像翩翩飛舞的蝴蝶，在熱帶季風林內，隨風搖曳生姿。蝴蝶蘭在台灣最早的採集紀錄是 1897 年英國人 Henry 在恆春半島採得，真正揚名於世界是在 1952 年得到國際蘭展的冠軍。自此之後，蝴蝶蘭就成了商人大肆蒐集的目標，經過多年的採集之後，搖曳生姿的蝴蝶蘭倩影，在天然林中已難得一見了。蝴蝶蘭主要分布於菲律賓和台灣，台灣是它的分布最北界，僅分布於大武、恆春半島東岸和蘭嶼的熱帶季風林內。

攝影／呂勝由

台灣一葉蘭 *Pleione bulbocodioides* (Tranch.) Rolfe

蘭科 Orchidaceae

附生蘭，植物體由一個大型的假球莖及一枚葉子構成，所以稱為一葉蘭。花頂生，1~2朵，花色豔麗，顏色變異頗大，由偏白的淡粉紅至深粉紅都有。台灣一葉蘭生育地屬本省盛行雲霧帶中之霧林帶及櫟林帶，屬演替初期的植物，生長發育需要充足的陽光，因此經常生長在垂直的岩壁上，春季開花。因其花姿非常優雅而享譽國際，多年來承受非常大的採集壓力，野外族群已大幅減少。

攝影／鄭育斌

大花羊耳蒜 *Liparis nigra* Seidenf.

蘭科 Orchidaceae

地生蘭，植株肉質多汁，花序頂生、直立，花多，深紫紅色，為台灣產花色最深，花瓣最大的羊耳蘭類植物。羊耳蘭屬（*Liparis*）全世界約 200 餘種，廣泛分布於全世界，台灣地區總共有 21 種，其中包含有地生蘭和附生蘭。羊耳草主要分布於全台低海拔山區林下，尤其北部山區非常常見，最早的採集紀錄是 1895 年英國人 Henry 採自淡水，並曾帶回 Kew Garden 栽植。

高山雛蘭 *Amitostigma alpestre* Fukuyama

蘭科 Orchidaceae

矮小的草本地生蘭，植株高僅約10餘公分，通常僅具兩片葉片。花2~3朵，淡紫色或粉紅色，白色的唇瓣基部呈三裂狀，上有粉紅色斑點。高山雛蘭是台灣高山的特有種，僅分布在中央山脈海拔3,000公尺以上的高山地區。1933年首度由福山伯明（Fukuyama）於南湖大山採得，並命名為新種。

攝影／鄭育斌

金線蓮 *Anoectochilus formosanus* Hayata

蘭科 Orchidaceae

珍貴的地生蘭，植株小巧可愛，高僅約5~10公分，開花期再加上花軸，也不過約20公分左右。名為金線蓮，主要是在絨毛狀的墨綠色葉片上，有白色的網紋。金線蓮廣泛分布在全台各地，多生長在潮濕的林下或石頭上。由於具有藥效，野外族群被大肆採集，再也難以看到大片的族群。金線蓮在台灣最早的採集紀錄是1895年英國人Henry在屏東萬金庄採得，1914年才由日本學者早田文藏氏（Hayata）命名。

攝影／呂勝由

台灣白及 *Bletilla formosana* (Hayata) Schltr. form. *formosana*

蘭科 Orchidaceae

地生蘭，扁壓狀的球莖埋於土中，葉片和禾草非常類似，不開花時不容易區分。花序頂生、纖細，花白至粉紅色。台灣白及喜歡生長在陽光充足的草生地或公路邊坡，廣泛分布在全台各地，從海岸邊至中高海拔均有，花期大約在5~8月。

攝影／呂勝由

台灣竹節蘭 *Appendicula formosana* Hayata

蘭科 Orchidaceae

懸垂性的附生蘭，植株密集叢生狀，附生於低海拔森林樹幹上。葉片兩列互生，花序側生於葉腋，花小型，白色。台灣竹節蘭是台灣特有種，主要產於恆春半島山區，向北可分布至台東太麻里。同屬另一種產於台灣的種類，長葉竹節蘭（*A. terrestris* Fukuyama）僅分布於蘭嶼。

攝影／鄭育斌

小鹿角蘭 *Ascocentrum pumilum* (Hayata) Schltr.

蘭科 Orchidaceae

小型的附生蘭，植物體矮小，葉厚革質，針狀，上表面中央有一縱溝。花腋生，總狀花序具3~10朵花，紫紅色，非常顯眼。小鹿角蘭是台灣特有種，分布於中、北部1,500~2,300公尺的山區森林內，通常附生於樹幹上，或是樹枝分枝凹處。鮮豔的花色是台灣原生蘭少見的。

攝影／呂勝由

台灣根節蘭 *Clanthe formosana* Rolfe

蘭科 Orchidaceae

根節蘭屬是台灣蘭科的大屬之一，總共有17個種，全都屬於地生蘭。台灣根節蘭是常見根節蘭之一，是台灣特有種，普遍分布於低海拔的山區的森林內。植株大型叢生狀，花序長，花色鮮黃。台灣根節蘭可作為季風雨林地生蘭的代表種。

攝影／鄭育斌

台灣喜普鞋蘭 *Cypripedium formosanum* Hayata

蘭科 Orchidaceae

地生蘭，根莖匍匐，分叉。莖直立，高約20公分。葉成雙，無柄，圓扇形，膜質。花大型、單朵，顏色非常鮮豔，由淡粉紅色到紫紅色，唇瓣呈囊袋狀。台灣喜普鞋蘭為台灣特有種，分布於海拔2,400公尺左右的山區，型態和日本喜普鞋蘭（*C. japonicum*）非常像，僅在花色上略有差異。台灣的四種野生喜普鞋蘭，都屬於高山性的野生蘭，除台灣喜普鞋蘭為特有種外，小喜普鞋蘭（*C. debile* Reichb. f）為東亞分佈，寶島喜普鞋蘭（*C. segawai* Masamune）是台灣特有種，奇萊喜普鞋蘭（*C. macranthum* Sw.）廣泛分布於世界各地，由歐洲經西伯利亞至日本都有分布，是屬於冰河孑遺物種。

攝影／呂勝由

赤色毛花蘭 *Eria tomentosiflora* Hayata

蘭科 Orchidaceae

又稱樹絨蘭，是一種懸垂性的附生蘭，常大片叢生，枝條呈圓柱型，末端膨大。花莖側生，花黃綠色至紅褐色，外披卷毛。本種大多生長在陰濕的樹幹上，或霧氣重的森林，分布範圍廣泛，全台各地低海拔山區都有分布。

攝影／呂勝由

美冠蘭 *Eulophia graminea* Lindl.

蘭科 Orchidaceae

又稱為禾草芋蘭，因為它的葉片類似於禾草，沒有開花的時候，並不容易分辨。美冠蘭是少數能生長於海岸沙灘的蘭花，粗大的卵型的假球莖，深埋在砂土中。葉線型至披針形，型態似禾草。花序總狀或圓錐狀，直立，高可達40公分，花暗綠色，具有褐色脈，鬆散的分布在花軸上，開花的時候無葉。美冠蘭廣泛分布於東亞地區和中南半島，在台灣地區大多分布於海岸沙地，但是最早採集到的標本卻是由川上龍瀰和森丑之助，於1908年採自能高山，可見美冠蘭的分布非常廣泛。

攝影／鄭育斌

台灣松蘭 *Gastrochilus formosanus* (Hayata) Hayata

蘭科 Orchidaceae

附生蘭，莖匍匐狀，細長，節處生根。葉兩列互生，肉質。花2~3朵及生於短花序上，花黃色或綠色，唇瓣囊袋狀，黃色，上有斑點。台灣的松蘭屬植物共有9種，台灣松蘭為最常見者，主要分布於海拔500~2,500公尺的山區，附生於樹幹上。

攝影／呂勝由

高山粉蝶蘭 *Platanthera sachalinensis* Fr. Schmidt

蘭科 Orchidaceae

粉蝶蘭屬是典型的高山蘭花，在台灣總共有 8 種 1 變種，主要都分布在高海拔山區的草原中。高山粉蝶蘭是一種生長於高山箭竹灌叢的小型蘭花，花序長約10~30公分，花多而密，花小型、白綠色。零星分布在海拔 2,300~2,800 公尺的高山箭竹灌叢。

攝影／呂勝由

紅小蝶蘭 *Ponerorchis kiraishiensis* (Hayata) Ohwi

蘭科 Orchidaceae

又稱為奇萊紅蘭，是高山的小草本蘭花，株高僅約10餘公分，具有卵型的地下塊莖。葉1~2片，基部包莖。花紅紫色，約1~2朵，唇瓣與蕊柱相連，粉紅色。最早由大橋捨三郎採自奇萊山海拔約 3,300 公尺山區，所以又稱為奇萊紅蘭。本種是台灣小蝶蘭屬中分布最廣的種類，通常分布在海拔 3,200 公尺以上的山區。除本種外，另還有三種小蝶蘭屬植物，都分布於高海拔山區。

攝影／呂勝由

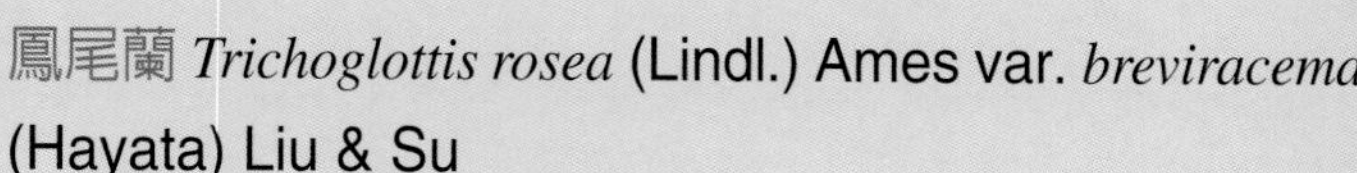

鳳尾蘭 *Trichoglottis rosea* (Lindl.) Ames var. *breviracema* (Hayata) Liu & Su

蘭科 Orchidaceae

附生蘭，莖懸垂叢生狀，葉帶狀，亮綠色，革質，兩列互生。花序軸短，約2~5朵花，長於各節與葉片相對位置。花白色，肉質，在唇瓣上有紫色斑塊。鳳尾蘭主要分在於南部山區，宜蘭蘇澳一帶也有分布，以往在南部低海拔山區森林內非常常見。

攝影／鄭育斌

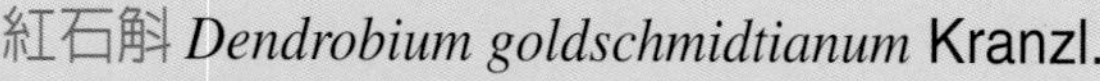

紅石斛 *Dendrobium goldschmidtianum* Kranzl.

蘭科 Orchidaceae

附生蘭，莖密集叢生，直立或懸垂，可長達 60 公分。莖節肥短，葉線狀長橢圓形。花序開於莖節處，落葉後開花。花密集，5~9朵，深紫紅色，具紫色脈紋。僅產於蘭嶼地區海拔約 200~400 公尺山區。

攝影／呂勝由

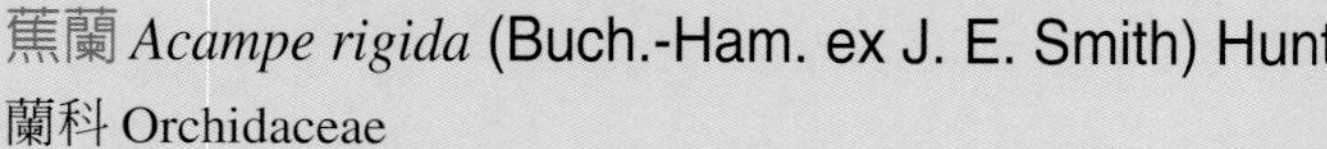

蕉蘭 *Acampe rigida* (Buch.-Ham. ex J. E. Smith) Hunt

蘭科 Orchidaceae

附生蘭，莖直立形，長可達1公尺以上。葉帶狀，兩列互生，革質。花莖側生，花肉質，有香味，花瓣黃色，有棕紅色橫斑。蕉蘭在台灣的分布很廣，主要產於向陽的溪谷石壁，是少見的岩生型附生蘭，常成群出現。

攝影／呂勝由

萬紫千紅

野生杜鵑花

Wild Azalea

◎撰文／呂勝由、曾彥學
◎Text ／ Sheng-you Lu
Yen-Hsueh Tseng

全世界的杜鵑花大約有900種左右，主要分布於北半球熱帶、亞熱帶至高山寒帶地區。台灣產15種杜鵑涵蓋了各式各樣的習性、分布、形態與花色，就生態習性而言，著生性的著生杜鵑、溪流岩岸的烏來杜鵑、成群出現的南澳杜鵑、台灣杜鵑、南湖杜鵑、玉山杜鵑、細葉杜鵑、紅毛杜鵑及台灣高山杜鵑等。就地理分布而言，烏來杜鵑僅產於台北縣境內北勢溪鸕鶿潭一帶的河岸；唐杜鵑則呈南北兩端分布；馬銀花及細葉杜鵑等主要分布在本島中部；長卵葉馬銀花則較常出現在南部高雄一帶山區。就外觀形態而言，著生杜鵑的花冠黃色、金毛杜鵑與唐杜鵑是磚紅色；台灣杜鵑、玉山杜鵑及南湖杜鵑等是鐘型花冠，西施花的花冠則是典型的漏斗型，顯現出台灣的杜鵑花多彩多姿的身影。

There are approximately 900 species of azalea in the world. They are primarily distributed in the tropics, semi-tropical areas and the alpine areas of the northern hemisphere. Taiwan produces 15 species of azalea, with varied habit, distribution, form and flower in terms of preferred habitat, the range is wide, from the epiphytic azaleas, the Wulai azaleas that live in creek cliffs, the Nan'ao azaleas, Taiwan azaleas, Nanhu azaleas, Yushan azaleas, Noriakianumi azaleas, Red hair azaleas and Taiwan alpine azaleas that grows in clumps. In terms of geographical distribution, The Wulai azalea is produced only on the banks of the Lutze Pond on the Peishi stream, and azaleas is distributed north and south of Taiwan; and noriakianumi azaleas are primarily found in the central part of the island of Taiwan; Rhododendron is more commonly seen in the south, in the area around Kaohsiung. Epiphytic azaleas is yellow, Gloden hair azaleas and azaleas are brick-red; formosa azaleas, Yushan azaleas and Nanhu azaleas are bell-shaped flowers, Taiwan azaleas has a characteristic funnel shape. All these variations go to show the wide variety of forms and types of azaleas to be seen in Taiwan.

■台灣高山杜鵑。

攝影／呂勝由

——談到杜鵑，便會想到陽明山的杜鵑花季。的確，假如有親自感受到萬紫千紅、百花爭豔的身境，就會瞭解為何杜鵑花是世界名花。杜鵑在中國最早記載於南北朝，大約五世紀左右，當時作為藥用。至唐朝時期大約八、九世紀，許多詩人便將杜鵑花作為詩詞歌賦的體裁來吟詠，如李白：蜀國曾聞子規鳥，宣城還見杜鵑花；一叫一迴腸一斷，三春三月憶三巴。之後歷代便有許多文獻流傳下來。至於當成園藝花卉則是近代的事。真正將它發揚光大，成為世界著名花卉，則是歐美各國的植物學家、園藝學家、採集家以及

■長卵葉馬銀花。 攝影／呂勝由

台灣原生杜鵑花的特色

南澳杜鵑（埔里杜鵑、毛柱杜鵑）
Rhododendron breviperulatum Hayata

杜鵑科 Ericaceae

常綠小灌木。葉半紙質，橢圓形，兩面披剛毛。花芽單一，頂生，每一花芽2~4朵；花冠漏斗狀，淡紫紅色，偶見白色品系。蒴果長橢圓形。喜生長於干擾過後的環境，如崩塌地或裸露地等。屬演替早期的先驅陽性植物。常見於台灣二葉松或台灣五葉松林下的灌木層。台灣特有種。主要分布台灣中央山脈中部兩側海拔400~2,200公尺地區，可見於台中、南投、嘉義、花蓮、宜蘭等縣。目前族群數量普通，族群結構穩定，但自然分布集中於中央山脈中部兩側山地，中橫公路又貫穿其中，因此生育地有隨時受人為干擾的風險存在。

西施花（阿里山杜鵑、青紫花）*Rhododendron ellipticum* Maxim.

杜鵑科 Ericaceae

常綠小喬木。小枝光滑。葉半革質，長橢圓形，兩面光滑，兩端銳形。花芽 2~5 個頂生，每一花芽僅一朵花，稀 2 朵；花冠漏斗狀，白至淡紅色。蒴果長橢圓形。常出現於低、中海拔闊葉林或針闊葉混交林邊緣，屬次優勢層耐陰性植物。廣泛種。分布中國、日本、琉球、台灣等地。台灣普遍分布於海拔200~2,400公尺地區，可見於台北、桃園、新竹、苗栗、台中、南投、嘉義、高雄、屏東、花蓮、宜蘭等縣。目前族群數量普通，族群結構穩定。

攝影／呂勝由

台灣杜鵑 *Rhododendron formosanum* Hemsl.

杜鵑科 Ericaceae

常綠小喬木。葉革質，披針狀長橢圓形，葉面光滑，葉背披灰白色貼伏狀毛茸。花芽頂生；花冠鐘形，白至淡紫紅色。蒴果長子彈形。常生長於低至中海拔的嶺脊或陡坡地區，且常常形成局部的純林，屬耐陰或半耐陰性植物。台灣特有種。分布台灣本島海拔600~2,400 公尺地區，可見於台北、桃園、新竹、苗栗、台中、南投、嘉義、高雄、屏東、花蓮、宜蘭等縣。目前族群數量尚豐富，族群結構穩定。

園藝商等。他們到處搜集，大量繁殖，培育新品種，使杜鵑花成為庭園花卉商品，行銷全世界。導致全世界無論公園、廣場、植物園、花園、學校等到處可見杜鵑花的蹤跡。陽明山公園內最常見的平戶杜鵑、皋月杜鵑、西洋杜鵑等，都是園藝栽培品種，而非台灣原生杜鵑。

全世界的杜鵑花大約有900種左右，隸屬於杜鵑科（Ericaceae），主要分布於北半球熱帶、亞熱帶至高山寒帶地區，其中地生型杜鵑以中國大陸雲貴高原一帶為主要的分布中心，沿喜馬拉雅山脈的尼泊爾、錫金、不丹、西藏、緬甸等地向外擴展逐漸演變成今日的分布。另一群著生型杜鵑則是以東印度群島，即印尼、馬來西亞、新幾內亞及附近島嶼為主要分布中心，再向外擴展。亞洲大約850幾種，中國就有500多種，是全世界產杜鵑花最多的國家；北美洲有24種；歐洲有9種；澳洲僅產1種；而非洲及南美洲則無杜鵑花。台灣正位於東亞地生型杜鵑與著生型杜鵑分布的交匯地區，加上島內氣候變化萬千，高山群峰層疊，地形極為複雜，故衍生許多野生杜鵑花。

有關台灣原生的杜鵑花，最早記載於Henry（1896）的「台灣植物名彙」當時僅記載金毛杜鵑

南湖杜鵑 *Rhododendron hyperythrum* Hay.

杜鵑科 Ericaceae

常綠小喬木。葉厚革質，長橢圓狀披針形，葉面光滑，葉背披黃褐色貼伏狀毛茸。花序頂生；花冠鐘形，花色白至淡紫紅。蒴果長子彈形。生長於中高海拔稜脊處、陡坡或林緣地。屬耐陰性植物。台灣特有種。分布海拔1,400~3,700公尺的山區，僅見於台中及花蓮兩縣境內的南湖大山、清水山、嵐山等高山。目前族群數量少，族群結構尚穩定，但因生育地狹隘，加上又是登山客活動地區，有受干擾之虞。

烏來杜鵑（柳葉杜鵑、金平氏杜鵑）

Rhododendron kanehirai Wilson

杜鵑科 Ericaceae

常綠小灌木。葉半紙質，線狀披針形，兩面披褐色剛毛。花芽單一，頂生，每一花芽2~4朵；花冠漏斗狀，有桃紅、粉紅、淡紫、紫紅等四種顏色。蒴果長橢圓形。常出現在河岸兩旁的岩石隙縫中，為典型的河岸杜鵑。屬陽性先驅植物。常與木槿、甜根子草、小葉桑等伴生。台灣特有種。原生育地僅分布於台灣北部新店溪上游的北勢溪，海拔100~300公尺的溪畔，以乾溝、鸕鶿潭一帶為主要分布中心。為台灣原生杜鵑植物中族群數量最少，分布最狹隘的一種。自從民國七十三年翡翠水庫竣工後，將其生育地全部淹沒。目前已無野外生育地，十幾年來皆無野外採集記錄，目前僅有少數移植栽培者。野外絕滅級（Extinct in the Wild, EW）。民國七十七年政府曾公告為法定珍貴稀有植物。

著生杜鵑（川上氏杜鵑）*Rhododendron kawakamii* Hayata

杜鵑科 Ericaceae

著生性常綠小灌木。葉半革質，長橢圓形，兩面光滑。花芽2~5個頂生，每一花芽僅一朵花，稀2朵；花冠漏斗狀，黃色。蒴果長橢圓形。常著生於中海拔闊葉林或針闊葉混交林優勢植物冠層枝幹上。被附著的樹木以台灣特有種紅檜為主。台灣普遍分布於海拔1,500~2,600公尺山地，可見於台北、桃園、新竹、苗栗、台中、南投、嘉義、高雄、屏東、花蓮、宜蘭等縣。目前族群數量普通，族群結構穩定。

攝影／呂勝由

（*Rhododendron oldhamii* Maxim.）及台灣杜鵑（*R. formusanum* Hemsl.）兩種，爾後陸續經過日據時代Hayata（1911-1921），Wilson（1925），Sasaki（1928），Suzuki（1935），Kanehira（1936），Masamune（1936），Ohwi（1937）；光復後李惠林（1963、1973），許建昌（1973），應紹舜（1976），楊遠波、呂勝由（1989）等專家學者精心研究，各學者對於台灣杜鵑花科記載約14種至29種之多。根據最近第二版台灣植物誌（李惠林、楊遠波、呂勝由、曾彥學1999）將台灣原生杜鵑處理為15種。

而這15種卻涵蓋各式各樣的形態與花色，且各具特色。就生態習性而言，著生杜鵑（*R. kawakamii* Hay.）是台灣原生杜鵑花中，唯一附生於別種植物體上的杜鵑，其他杜鵑則地生，而它也僅出現在本島中海拔的霧林帶；烏來杜鵑（*R.kanehirai* Wilson）僅出現在溪流的岩岸，是台灣原生杜鵑花中，唯一屬於亞熱帶河岸杜鵑者；喜歡成群結隊一大片出現的杜鵑有南澳杜鵑（埔里杜鵑）（*R.brevip ralatam* Hemsl. & Wilson）、台灣杜鵑、南湖杜鵑（*R.hyperyfhrum* Hay.）、玉山杜鵑（*R.pseudochrysanthum*

守城滿山紅（馬里士杜鵑）

Rhododendron mariesii Hemsl. & Wilson

杜鵑科 Ericaceae

常綠小灌木。葉半紙質，橢圓形，兩面披剛毛。花芽單一，頂生，每一花芽2~4朵；花冠漏斗狀，淡紫紅色，偶見白色品系。蒴果長橢圓形。陽性植物，於土壤及空氣潤濕的環境中生長較為良好，其生育地於七星山一帶者，為灌草叢之陡坡居多，於坪林一帶者多生長於河谷兩岸，鴛鴦湖者生長於沼澤地邊緣地區頗為茂盛，為濕地草本植被演替至山地中生森林內（檜木林）的過度灌叢植被。廣泛種。分布中國及台灣等地。台灣分布於海拔200~1,800公尺地區，可見於宜蘭、台北、台中、南投、屏東等縣。目前族群數量普通，族群結構穩定。

細葉杜鵑（志佳陽杜鵑、北部高山紅花杜鵑）

Rhododendron noriakianumi T. Suzuki

杜鵑科 Ericaceae

常綠小灌木。葉半紙質，橢圓形，兩面披剛毛。花芽單一，頂生，每一花芽2~4朵；花冠漏斗狀，淡紫紅色，偶見白色品系。蒴果長橢圓形。在原生杜鵑花中葉2片最細小者。常出現於中、高海拔開闊草生地與玉山箭竹混生或二葉松林下與高山芒混生，屬陽性植物。台灣特有種。分布於本島中北部海拔1,500~2,800公尺地區，可見於新竹、台中、南投等縣。目前族群數量普通，族群結構穩定。

金毛杜鵑 *Rhododendron oldhamii* Maxim.

杜鵑科 Ericaceae

常綠小灌木。葉半紙質，橢圓形，兩面披剛毛。花芽單一，頂生，每一花芽2~4朵；花冠漏斗狀，磚紅色。蒴果長橢圓形。先驅陽性植物，適應力強，尤耐瘠薄的土壤，多見於林道，產業道路兩旁邊坡、火燒跡地、崩塌地及河谷壁等，常形成灌木草叢之植被地區。台灣特有種。分布於本島低海拔至2,500公尺地區，可見於宜蘭、台北、新竹、台中、南投、嘉義、高雄、屏東等縣。目前族群數量普通，族群結構穩定。

攝影／呂勝由

Hay.）、細葉杜鵑（*R.noriakianumi* T.Suzuki）、紅毛杜鵑（*R.rubropilasum* Hay.）及台灣高山杜鵑（*R.rubropilo* Hay.Var. *baiwenalpi* (ohwi) Lu, Yang&Tseng）等，每當花季來臨，百花爭放，場面頗為壯觀；南澳杜鵑常盤據低海拔的二葉松林下，中興大學惠蓀林場的杜鵑嶺是最佳觀賞地方。台灣杜鵑是台灣原生杜鵑花中，唯一可形成森林的杜鵑，台灣大學溪頭遊樂區的鳳凰山是較易觀賞的地點；玉山杜鵑可適應經年強風吹襲、冬雪削壓的環境，是台灣分布海拔最高的杜鵑，松雪樓後的合歡東峰是觀賞玉山杜鵑不可錯過的

■馬銀花。 攝影／呂勝由

馬銀花 *Rhododendron ovatum* Planch

杜鵑科 Ericaceae

常綠小灌木。葉半紙質，橢圓形，兩面披剛毛。花芽單一，頂生，每一花芽2~4朵；花冠漏斗狀，淡紫紅色，偶見白色品系。蒴果長橢圓形。生長於河谷兩岸陡坡，屬陽性偏中生之植物，其生育地主要植被有台灣二葉松、台灣五葉松、台灣肖楠、威氏帝杉、少量紅檜及一些楓香、櫸木、阿里山千金榆、圓果青剛櫟等針闊葉混生地區，於土壤略濕潤地區生長較為良好。廣泛種。分布中國、台灣等地。台灣僅分布於中部及北部山地，僅見於桃園、新竹、台中等縣。目前族群數量普通，族群結構穩定。

長卵葉馬銀花 *Rhododendron ovatum* Planch var. *lamprophyllum* (Hayata) Y.C.Liu, F.Y.Lu, & C.H.Ou

杜鵑科 Ericaceae

常綠小喬木。小枝光滑。葉半革質，長橢圓形，兩面光滑。花芽2~5個頂生，每一花芽僅一朵花，稀2朵；花冠漏斗狀，白至淡紅色。蒴果長橢圓形。常出現於低、中海拔闊葉林或針闊葉混交林邊緣，屬下層耐陰性植物。台灣特有變種。分布於中部及南部海拔1,700~2,200公尺地區，可見於南投、高雄等縣。目前族群數量不多，族群結構不穩定，且因生育地狹隘，加上又是登山客活動地區，有受干擾之虞。瀕危級（Endangered, EN）。

玉山杜鵑（森氏杜鵑、紅星杜鵑）

Rhododendron pseudochrysanthum Hayata

杜鵑科 Ericaceae

常綠小灌木。葉革質，橢圓形，葉面光滑，葉背幼時披密毛茸，成熟時則光滑。花芽單一頂生；花冠鐘形，白至淡紫紅色。蒴果長子彈形。主要生長於高海拔的林緣、稜脊或陡坡的灌叢地區，常與玉山圓柏或玉山箭竹形成優勢的植物社會，屬耐陰或半耐陰性植物。台灣特有種。分布台灣本島海拔2,800公尺以上高山，可見於新竹、苗栗、台中、南投、嘉義、高雄、花蓮等縣。目前族群數量尚豐富，族群結構穩定。

攝影／呂勝由

景點；南湖杜鵑則常出現在石灰岩的環境中，南湖大山前的五岩峰最為壯觀，只是必須花費兩天的路程，才能目睹真顏；梨山附近的二葉松林下則是細葉杜鵑形成優勢；至於台灣高海拔地區的二葉松林下，則是由紅毛杜鵑與其變種台灣高山杜鵑稱霸，中橫霧社支線鳶峰休息站，可發現紅毛杜鵑的芳蹤。

就地理分布而言，烏來杜鵑僅產於台北縣境內北勢溪鸕鶿潭一帶的河岸；唐杜鵑（*R.simsii* Planch）則呈南北兩端分布；馬銀花（*R.ovatum* Planch）及細葉杜鵑等主要分布在本島中部；長卵葉馬銀花（*R.ovatum* Planch var. *lamprophylum* (Hay.) Liu, Lu & Ou）則較常出現在南部高雄一帶山區。就世界植物地理而言，著生杜鵑可能是東印度群島著生型杜鵑分布的北界。就外觀形態而言，著生杜鵑的花冠黃色，也是與眾不同；金毛杜鵑與唐杜鵑的花冠則是磚紅色；細葉杜鵑是台灣原生杜鵑花中葉片最小者；花冠鐘型者有台灣杜鵑、玉山杜鵑及南湖杜鵑等；西施花（*R.ellipticum* Maxim.）的花冠則是典型的漏斗型。

紅毛杜鵑 *Rhododendron rubropilosum* Hayata

杜鵑科 Ericaceae

常綠小灌木。葉半紙質，橢圓形，兩面披剛毛。花芽單一，頂生，每一花芽2~4朵；花冠漏斗狀，淡紫紅色，偶見白色品系。蒴果長橢圓形。喜生長於開闊及台灣二葉松、華山松的疏林中，常與高山芒、玉山箭竹及巒大蕨等混生，屬先驅性的陽性植物，亦為中、高海拔山區火災適存之植被。台灣特有種。台灣普遍分布於中部海拔1,000~3,300公尺山區，可見於新竹、苗栗、台中、南投、嘉義、宜蘭等縣。目前族群數量普通，族群結構穩定。

台灣高山杜鵑 *Rhododendron rubropilosum* Hay. var. *taiwanalpinum* (Ohwi) S.Y.Lu, Y.P.Yang & Y.H.Tseng

杜鵑科 Ericaceae

常綠小灌木。葉半紙質，橢圓形，兩面披剛毛，葉緣反捲。花芽單一，頂生，每一花芽2~4朵；花冠漏斗狀，淡紫紅色，偶見白色品系。蒴果長橢圓形。常出現於中、高海拔開闊草生地與玉山箭竹混生或二葉松林下與高山芒混生，屬陽性植物。台灣特有變種。台灣分布於北部海拔2,800~3,000公尺的高山地區，可見於台北、桃園、新竹、苗栗等縣。目前族群數量普通，族群結構穩定。

唐杜鵑（大屯杜鵑、中原氏杜鵑）*Rhododendron simsii* Planch

杜鵑科 Ericaceae

常綠小灌木。葉半紙質，橢圓形，兩面披剛毛。花芽單一，頂生，每一花芽2~4朵；花冠漏斗狀，磚紅色。蒴果長橢圓形。常出現於低、中海拔闊葉林或針闊葉混交林邊緣，屬次優勢層耐陰性植物。生長於亞熱帶以至熱帶低山地區的杜鵑，屬不耐陰之中生植物，多見於陡坡、稜脊之林緣灌草叢或不太遮蔽的疏林中。目前族群數量普通，族群結構穩定。

攝影／呂勝由

空中的居民

附生植物

Epiphyte

◎撰文／徐嘉君 邱文良
◎ Text／Chia-Chun Hsu
Wen-Laing Chiou

所謂的附生植物，其基本定義是指某一群植物，生長在其它的植物上，但並不吸取被附生的植物的養份進而影響宿主（被附生植物生長其上的植物稱之爲宿主。）生長，而宿主除了提供生長環境外，也不影響附生植物。

■多采多姿的附生植物擴展了森林的生物多樣性。

除了蘚苔與地衣等較原始的非維管束附生植物，全世界主要有三大類的維管束附生植物。第一類為鳳梨科的植物，它們主要分布在中南美洲的熱帶雨林；第二類為蘭科植物；第三類則為蕨類植物，後兩者都是廣泛分布全世界的植物。

附生植物生長在空中，所面臨的問題之一是乾旱，因為這類植物無法接觸土壤，唯一的水份來源是雨水及霧水。雖然有些地區的生育地雨量豐沛，然而儘管經常下雨，但在兩場雨之間，由於樹幹上並無土壤，無法貯存水分，因此環境十分乾旱，這是附生植物所必須適應的環境。

蕨類植物亦是維管束附生植物中三大分類群之一。 攝影／徐嘉君

崖薑蕨是鳥巢型的大型附生蕨類。 攝影／徐嘉君

攝影／蕨類研究室

■附生植物在型態及生理上已發展出適應在樹冠層生存的機制。
攝影／徐嘉君

許多附生植物衍生出「CAM」(Crassulacean Acid Metabolism)的光合作用方式來適應乾旱環境，CAM型植物普遍生長在乾旱地區，生長速度十分緩慢，因為它們最優先考量的是如何保存水份，而非快速生長。大部分地球上具充足水份地區的植物光合作用形式多為C3或C4型，而非CAM型。（植物的光合作用以其代謝程序的不同分為C3 、 C4及CAM三種形式。）

綠色植物需要陽光將二氧化碳轉為醣類，以獲取能量，所以以C3 、 C4形式光合作用的植物，在白天有陽光時將氣孔打開，以吸收二氧化碳，並製造醣類。相反的，CAM型的植物白天則將氣孔緊閉，僅捕捉並貯存太陽能，等到晚上才打開氣孔。

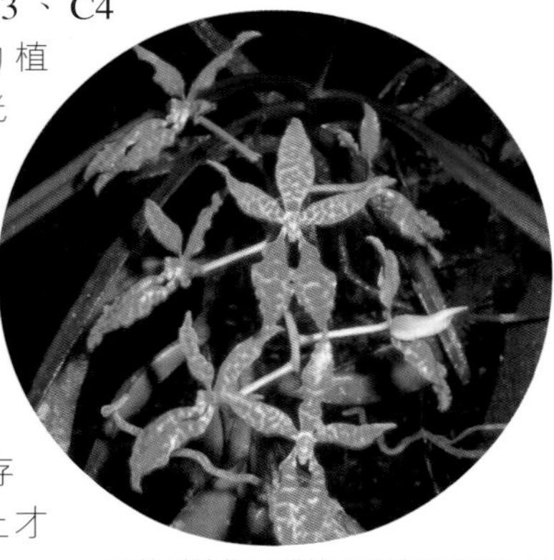
■蘭科植物是維管束附生植物中三大分類群之一。
攝影／徐嘉君

■樹橋是研究附生植物十分方便的設施。 攝影／徐嘉君

植物界的吸血鬼

寄生植物

Vampires of Plants — Parasitic Plants

◎撰文／邱少婷

◎Text／Shau-Ting Chiu

寄生植物以吸器穿透寄主植物的保護層，掠奪養分或水分，是植物界的「吸血鬼」。台灣以桑寄生科莖寄生植物的種類最多，其次爲蛇菰科和列當科的根寄生植物種類也不少。除了檀香爲栽培喬木，包括原生灌木、草本或藤本植物，共8科14屬34種以上，大部份分布於中低海拔區域，以低海拔開發造成生育地的縮減，處於受威脅或瀕臨絕滅危機的種類包括菱形奴草、海桐生蛇菰、寄生鱗葉草、刀葉槲寄生等。寄生植物在生態系中除了是掠奪者，但在食物網串連中也可能是其他物種的食物來源，以往因危害經濟林、行道樹、植被景觀等，而被視爲有害生物加以清除，可能因此造成連鎖反應而影響其他物種的生存。當寄生植物與人類主觀的利益衝突而被部分曲解或忽視的生態角色，將威脅生物多樣性保育與永續經營的自然資源平衡，是不可不深思的議題。

Parasitic plants invade the protective layers of their host plants with haustoria and plunder their hosts' nutrients and water. They are the "vampires" of the plant kingdom. The majority of parasitic species in Taiwan is the stem-parasite Loranthaceae, followed by the root-parasite Balanophoraceae and Orobanchaceae. Besides the cultivated trees of sandalwoods, there are eight families, fourteen genera and 34 species. They include the native shrubs, herbs and vines. Most of them distributed in the middle- and lower-altitude areas. Due to the development causing the decline of habitats in the low altitudes, species that are endangered or facing the crisis of extinction include the *Mitrastemon kanehirai* Yamamoto, *Balanophora tobiracola* Makino, *Epirixanthes elongata* Bl., and *Viscum multinerve* (Hayata) Hayata. Parasitic plants are not only the plunderers in the ecosystem, they also can provide a source of food for other links in the food chains. Because they usually damaged the artificial forests, street trees and ornamental landscape, they are regarded as a menace and cleared away. These elimination may arouse a chain reaction with the impact to the existence of other living organisms. When the benefits of parasitic plants conflict with the interests of humans, the natural function of these plants is twisted or ignored. It will menace biodiversity for conservation and sustainability in the management of natural resources. We must considerate these interrelated issues carefully and deeply.

■森林的吸血鬼－寄生植物。　攝影／邱少婷

植物界的「吸血鬼」

大部分植物捕捉光能，行光合作用，合成養分，是自然界的生產者。但某些植物以吸器穿透寄主植物的保護層，盜取養分或水分，嚴重危害寄主，也就是植物界的「吸血鬼」，一般所稱的寄生植物。

它們必須仰賴寄主才能生存，以往因危害經濟林、行道樹、植被景觀等，而被視為有害生物加以清除。但生態系中食物網環環相扣，牽一髮而動全身，寄生植物與人類主觀的利益衝突而被部分曲解或忽視的生態角色，將威脅生物多樣性保育與永續經營的自然資源平衡。

根據寄生植物之特性，依其寄主被寄生的部位可分根寄生（如：列當）和莖寄生（如：桑寄生科），依其營養體主要在寄主內或外，可分內寄生（如：蛇菇科）和外寄生（如：無根草），依其對寄主營養掠奪程度可分全寄生(如：菟絲子，無法行光合作用自營生活)和半寄生(如：桑寄生科，具葉綠體等行光合作用)。有的寄生植物僅在幼年期必須仰賴寄主，成株則可獨立生存，如：檀香(*Santalum album* L.)。有的寄生植物在寄主植物體內形成寄生營養體的生長擴展，即所謂內生植物體系統（endophytic system），甚至無性萌芽形成同源體系，如：檜葉寄生。這些僅顯示寄生植物的多樣性，其生活史的複雜度更不在言下。

■杜鵑寄生(*Taxillus rhododendricolus*)在櫻樹上。 攝影／邱少婷

台灣的寄生植物以雙子葉植物為例，共8科14屬34種以上，除了檀香為栽培喬木，其他為原生灌木、草本或藤本植物，被列為稀有的種類包括：菱形奴草、海桐生蛇菇、寄生鱗葉草、刀葉槲寄生等，這些物種族群稀少，處於受威脅或瀕臨絕滅的危機中，但生存和繁殖機制的研究卻相當有限。

■桑寄生的綠葉襯紅花，特別吸引鳥的青睞。 攝影／邱少婷

啄花鳥、粉蝶、小灰蝶的最愛

■蓮花池寄生。 攝影／邱少婷

桑寄生類植物一直被視為經濟林業、果園、行道樹及植被景觀的危害生物之一，經過森林資源調查的深入了解，其實桑寄生植物的特殊演化，與啄花鳥類和一些粉蝶、小灰蝶的生活中食物來源習習相關，或許人類之恨，卻是啄花鳥、粉蝶、小灰蝶的最愛。

■蓮花池寄生的二級根吸器及萌蘗芽。 攝影／邱少婷

■寄生在裸子植物上。 攝影／邱少婷

目前台灣產桑寄生植物共4屬17種及2變種，大多數以啄花鳥類為傳播媒介。因桑寄生類的傳播與啄花鳥類的食性密切關係，其分布及物候性也相互影響。其生活史簡單可分以下數個階段：(1)種子黏著於寄主(2)種子萌發(3)確立感染機制：包括吸器侵入和吸收寄主的水、養分等，有的種類發展氣生根，並於寄主接觸發展二級吸器，強化感染效果。(4)繁殖與傳播：桑寄生的開花結果與傳播媒介（啄花鳥）的食性和排便行為息息相關，進而影響其傳播和分布。

以低海拔之蓮花池寄生（*Taxillus tsaii* Chiu）和中海拔之杜鵑寄生（*T. rhododendricolus* (Hayata) Danser）為例，蓮花池寄生3月至7月開花期，6月至11月熟果期，主要以綠啄花（*Dicaeum concolor*）為傳播鳥，也有紅胸啄花（*D. ignipectus*）之參與。杜鵑寄生9月至10月開花，10月至12月果熟期，主要以紅胸啄花為傳播鳥。種子萌發時，蓮花池寄生則以子葉及初生葉迅速成長為主，兼以突破寄主樹皮侵入；杜鵑寄生則以突破寄主樹皮侵入為先，一般可明顯看到寄主樹皮開裂現象，而後子葉及初生葉才萌發生長。因海拔溫差與生長季長短差異，桑寄生植物之生活史與其物候性及分布演化甚密。

■台灣最小的桑寄生植物－檜葉寄生。 攝影／邱少婷

■桑寄生植物的重要傳媒－綠啄花。 攝影／邱少婷

■看我如何根深蒂固－檜葉寄生侵入樹枝木質部。 攝影／邱少婷

除了在溫帶的純林區域或樹種較單純的森林，因桑寄生植物的寄主選擇有限才造成寄主單純化的現象，大部份桑寄生科植物不具寄主單種專一性。台灣中低海拔區域樹木物種豐富，因此桑寄生科植物的寄主專一性非常不明顯。侏儒型桑寄生植物—檜葉寄生（*Korthalsella japonica* (Thunb.) Engler），個體雖小，分布卻由南到北，低海拔到中高海拔，均可能發現。在台灣的十多種桑寄生植物中，僅松寄生（*Taxillus matsudai* (Hayata) Danser）主要以松科、裸子植物為主要寄主，高氏桯寄生（*Loranthus kaoi* (Chao) Kiu）以其他桑寄生植物為寄生對象，呈現重寄生現象（hyperparasitism），大部份的寄主記錄從裸子植物到被子植物、喬木、灌木到藤本，呈現多樣性。

台灣槲寄生（*Viscum alniformosanae* Hayata）是槲寄生屬在中海拔中陽性環境的普遍分布種類，但原為華南低海拔常綠闊葉林分布，且模式標本採自台灣的刀葉槲寄生（*Viscum multinerve* (Hayata) Hayata）卻有逐漸減少的現象，其寄主主要為殼斗科和樟科植物，目前台灣新採集的寄主記錄僅桂花（*Osmanthus fragrans* Lour.）、紅皮（*Styrax suberifolia* Hook. & Arn.）、三斗石櫟（*Pasania hancei* (Benth.) Schott. var. *ternaticupula* (Hay.) Liao）、白校欑（*Castanopsis carlesii* (Hemsl.) Hayata var. *sessilis* Nakai），且多為單株，族群數量極低，相距又太遠。這顯示低海拔過度開發、森林退縮過程、物種遷就環境演替，生物多樣性也在漸漸式微。

■族群漸稀少的刀葉寄生。

攝影／邱少婷

■森林聖誕節的裝飾。

攝影／邱少婷

■稀有的菱形奴草－像古老日本奴隸的光頭。 攝影／邱少婷

樹根的仙環

寄生植物以亞熱帶、熱帶分布的大花草科植物特別受矚目，亞洲熱帶雨林分布的大王花屬（*Rafflesia*）植物因包含目前世界上發現最大的花朵記錄，而成為觀光自然資源的保護重點。由於大王花生長在陰暗的熱帶雨林中，無根莖葉構造，種子萌發直接侵入葡萄科的蔓藤莖內，直到在寄主內蔓延成熟後，形成特大單性花朵的構造冒出寄主表皮外，才會被發現。授粉和生殖機制的研究相當有限，因其肉肝色的花朵中央柱散發腐臭氣味，推斷吸引蠅類授粉。但其特異的內部構造和習性，加上熱帶雨林的複雜性，至今傳播機制仍無法完全了解，更遑論精確的保育措施。

台灣雖沒有大王花，但卻有此科另一群會在寄主根系範圍產生「仙環」的奴草屬（*Mitrastemon*），是大花草科植物與Hydnoraceae親緣最近的橋樑，此群植物為全寄生且內寄生植物，兩性花、子房上位、雄蕊先熟，靠蜂類和蠅類傳粉，可能由鳥類做長距離的種子傳播。

台灣的兩種奴草屬植物：菱形奴草（*M. kanehirai* Yamamoto）和台灣奴草（*M. kawasasakii* Hayata）花期和地理分布區隔明顯，其中台灣奴草分佈在中低海拔闊葉林中，寄生於多種殼斗科植物的根部，花期在1~4月；菱形奴草目前僅發現於南投縣魚池鄉低海拔闊葉林中，唯一寄主記錄為白校欑，花期在9~12月。

由於菱形奴草的族群小且族群間遺傳變異程度低，似乎呈現稀有物種的特性，但族群內個體間卻有非常顯著的差異，這代表傳播或是族群擴展的限制性，有待深入的研究，才能提出更妥善的保育策略。

深藏不露

蛇菇科主要為內寄生且全寄生的根寄生植物，不到開花結果，不容易發現它們的蹤影。由於其花序形成肉質菇狀形態，從地下冒出時，遇到硬物障礙可能彎曲蛇行，故稱蛇菇。常被誤解為真菌類，而實際為開花植物。穗花蛇菇（*Balanophora spicata* Hayata）就是愛吸樹木腳趾頭的根寄生植物之一，廣泛分布於中、低海拔森林中。

在台灣眾多寄生植物中，只分布在特定狹窄地點除了菱形奴草，還有蛇菇科的海桐生蛇菇（*Balanophora tobiracola* Makino）和遠志科的寄生鱗葉草（*Epirixanthes elongata* Bl.）。海桐生蛇菇僅分布於花蓮低海拔林區，寄生鱗葉草分布於墾丁半島低海拔潮濕森林生態保護區，由於花期很短，以往很少被發現。

■愛吸樹木腳指頭的根寄生－穗花蛇菇。 攝影／邱少婷

鬼見愁

某些寄生植物只找特定的「倒楣鬼」寄主，當其終身巧取豪奪的對象，這些寄主遇到它們當然要發愁了。例如列當（*Orobanche caerulescens* Stephan *ex* Willd.），一種專門寄生在茵陳蒿根部的肥厚無葉綠體寄生草本，主要分布於中北部海岸。野菰（*Aeginetia indica* L.）專喜寄生在芒草和甘蔗等禾本科植物的根部，屬於低海拔廣泛分布種類。假野菰（*Christisonia sinensis* G. Beck）則分布在北

■最愛芒草根的野菰。

攝影／邱少婷

■鬼見愁－列當專門寄生在茵陳蒿的根部。

攝影／楊國禎

部山區海拔兩千多公尺的玉山箭竹根部。川上氏肉蓯蓉（*Boschniakia kawakamii* Hayata）則特喜分布於高海拔的玉山杜鵑根部。

攀緣吸附的軟骨頭

某些寄生植物不僅掠奪寄主的養份，還是攀緣吸附的蔓藤性植物，造成寄主嚴重的生理及重量負擔。主要的代表是廣泛分布於低海拔的菟絲子（*Cuscuta australis* R. Br.）及近海邊的無根草（*Cassytha filiformis* L.）。

寄主一但被「釘」上，必然邁入萬劫不復的命運。這些蔓藤性的寄生植物體，快速成長、擴展，吸盤按步就班地緊緊扣住養份來源，不到寄主乾枯力竭，絕不罷休。而且毫不挑嘴，來者不拒，因此，菟絲子或無根草走過之處必留下一片枯黃。

曾經將菟絲子暴曬烘乾五天五夜之後，黑色焦屍居然於數天後遇水則發，尖端露出黃色新生的嫩芽，可見菟絲子的「雜草性」（weedy）。傳聞標本館中溼度不夠低，菟絲子臘葉標本也可能像九命怪貓起死回生，在台紙上甦醒過來，慶幸的是生長一段時間後，因沒有活寄主可提供養分，終於結束這段神鬼傳奇。

■菟絲子－植物吸血鬼中的軟骨頭。 攝影／邱少婷

■分布高海拔，最愛玉山杜鵑根的川上氏肉蓯蓉。 攝影／邱少婷

■樟科植物中的異類－無根草，外寄生型。 攝影／邱少婷

超越顛峰的挑戰

高海拔

The High Altitudes

◎撰文／郭城孟 ◎ Text ／ Chen-Meng Kuo

攝影／蕨類研究室

翠池地區因較避風，圓柏長成直立的大樹。攝影／蕨類研究室

位於亞熱帶地區的台灣，因具有海拔3,000公尺以上的高山，而海拔高度每上升1,000公尺，平均氣溫降低大約6℃，因此在台灣海拔3,600公尺的地方，氣溫比平地低約20℃，是終年處在低溫的狀況，在南湖、雪山、玉山的山頭冬天還可看到積雪，因此也產生了與北方針葉林及極圈地區近似的景觀，包括寒原地區的碎石坡短草地、圓柏灌叢、杜鵑灌叢，以及北方針葉林型的冷杉林、鐵杉林、雲杉林等。

台灣的高山寒原（alpine tundra）與極地寒原（arctic tundra）最大的差異在於日照

Taiwan is positioned in a subtropical location, for which reason, in her mountains of over 3,000 meters for each increase of 1, 000 meters in elevation, there is an average drop in temperature of 6℃. As a result, in the areas of 3,600 meters above sea level, the air temperature is some 20 ℃ colder than on the plains. In these year-round conditions of low temperatures, the mountain peaks of Nanhu, Hsuehshan and Jade Mountain sometimes see snow accumulations, for which reason they have a vista much like that of the conifer forests and arctic climes, including the short grasses seen on gravelly arctic slopes, Juniperus squamata, and shrub, as well as boreal forests of fir, hemlock and Picea Spruce.

■雪山主峰，可見碎石坡上的圓柏矮盤灌叢。

週期和日照角度，極地寒原每天的日照長度變化很大，具有永晝和永夜的現象，不像高山寒原祇有幾個小時的差別，此外，台灣高山寒原接近北回歸線，因此的日照角度僅略為斜射，不像極地寒原是與地面至少66°以上的斜角，因此在台灣高山地區日照與否造成的溫差頗大，雖然因冰河北退的影響，生物種類分別退到高緯度及高海拔地區，因此高山寒原與極地寒原具有類似的生物，但是高山寒原的種類比極地寒原的要更耐溫差變化。

■ 雪山圈谷地區的高山寒原景觀。 攝影／蕨類研究室

攝影／蕨類研究室

The greatest difference in Taiwan's alpine tundra and arctic tundra lies in the cycle of sunlight and the angle of the sun. The arctic tundra sees a huge variation in the amount of sunlight received daily, with phenomena of the "midnight sun" and 24 hours of darkness. This is quite unlike the alpine tundra, where there are only a few hours' difference. Moreover, Taiwan's alpine tundra is near the Tropic of Cancer, so that the angle of the sun's rays is only slight, unlike the angles of at least 66 degrees seen in the arctic tundra areas. This means that the temperature variations caused by the changes in sunlight in Taiwan's alpine tundra regions require the plants of the alpine tundra region to be more tolerant to changes in temperature than those in the arctic tundra region. But because of the withdrawal of the glaciers northward, plant species retreated to higher latitudes and higher altitudes, making many of the species of the alpine tundra and arctic tundra similar.

The areas in Taiwan above 3,500 meters are mostly gravelly slopes. The slope of these hills, plus the strength of the wind, and the scarcity of soil added to the poor water retention qualities of the environment mean that the environment for plant life is very dry. The plants that live here are generally very short. The root portions delve deeply into fissures in the rock, and their above-ground parts are mostly stuck to the surface of the ground, reducing the area exposed to the wind. The leaf surfaces of these plants are very small, or the leaves are needle-shaped, in order to reduce the surface from which water could evaporate. Some are fleshy, with sunken stomata, and long fuzz or gills on their leaves, all the variations designed to reduce water loss and to enable the plants to survive when they are covered with snow. The *Epilobiuim nankotaizanense* Yamamoto, endemic species to Taiwan, is a good example. In autumn and winter, some of these plants dry out above ground, leaving only their underground portions living. Some plant species dry up completely come autumn, and pass the winter in the form of seeds. The mosses and lichens are the stars here.

台灣海拔3,500公尺以上的地區多形成碎石坡環境，加上地形陡峭，風勢強勁，土壤堆積極少，且水分保存不易，對生物而言是極為嚴酷的環境，因此這裡的植物都是低矮型的。植物根部常深入岩縫之中，地上部分也多緊貼地面生長，減少受風面積，植物體的葉面則變小或呈針狀，以減少蒸散面積，有的則形成肉質狀、氣孔下陷、葉面長著絨毛或鱗片...等，以減少水分的散失，並使其在降雪季節受大雪覆蓋仍能生存，台灣的特有植物南湖柳葉菜（*Epilobium nankotaizanense* Yamamoto）即是。部份植物到了秋冬季節，地上部分即行枯死，僅保留地下部分過冬，有些植物則是到了冬天時全株枯死，而以種子的形態過冬，此外苔蘚、地衣也是這裡的主角。

在碎石坡下緣緊鄰森林界線的地區，則是形成矮盤狀的圓柏（*Juniperus squamata* Lamb. var. *morrisonicola* (Hayata) Li & Keng）灌叢或杜鵑灌叢，圓柏灌叢主要分布在較陰濕的環境，而杜鵑灌叢則是在向陽開闊地，在中等環境可見二者混生。矮盤灌叢生長的地方較草生地而言，略為避風，所以植物體高可達1~2公尺，但仍呈緊密糾結的結構，以增強對風的抵受力。隨著離山頂稜線距離愈遠，灌叢植株高度逐漸增加，彼此的間距也增大，在接

On the gravelly slopes near the forest line, are clumps of *Juniperus squamata* Lamb. var. *morrisonicola* (Hayata) Li & Keng., or azalea shrub. The Juniperus are generally found in darker, moister areas, and azalea shrub in the more open areas. In areas whose conditions are between these two, both species can be seen mixed together. Short shrubs grow in areas that have more protection from the wind than grass areas, so the plants can reach 1 to 2 meters in height. but they still display a very dense structure, in order to increase their resistance to wind. As we move farther below the peak toward the forest line, the height of the shrubs gradually increases, while the distance between them also increases. As we near the forest line, the plants are close together and gradually change into a thicket. Moving upward, they exist as individual plants. Below the shrub thickets is the forest line. The upper limitation of the shrub thickets is the upper limit of the woody plants. This is why it is called the tree line. The belt of shrubs occurs between the forest line and the tree line, which makes them the link between the forest and the grassy areas. Called the "timber line", the forest line is actually not horizontal at all, but wanders up and down primarily influenced by the shape of the mountain in question. The elevation of the forest line is lower on a mountain ridge, and on hollows in the mountains, this line may move upward to a higher altitude. The reason is that the mountain ridge environment is more similar to an high mountain peak, which impacts the way plants grow. At the higher elevation mountain hollows, the environmental factors that in-

■碎石中的南湖柳葉菜。

攝影／蕨類研究室

近森林界線處的灌叢，植株多少相鄰，往上逐漸變成幾株一叢地散佈，再往上則呈單株散生。在矮盤灌叢的下限即是「森林界限（forest line）」，矮盤灌叢上限即是木本植物的盡頭，所以稱為「樹木界限（tree line）」，而矮盤灌叢帶即是「森林界限」與「樹木界限」的區域，也就是森林和草生地的推移帶，稱為「林木界線（timber line）」。森林界限並不是水平的，而是呈犬牙交錯狀的，主要是受到山脊與山窪的影響，森林線在山脊一帶的海拔高度較低，而在山窪則可向上發展至較高海拔的地方，原因是山脊一帶的環境因子比較類似高山山頂影響植物生長的因子，而在山窪地的環境因子比較類似較低海拔森林地區的影響因子。森林界線受環境因子的影響，主要是溫度，因此界線的海拔高度因地而異，在

■鐵杉。

攝影／蕨類研究室

fluence plant are more like those of the lower-elevation forest area. The primary environmental factor that impacts the forest line is temperature. Consequently, the altitude of this line varies with temperature. In the Swiss Alps, the forest line is around 2,000 m, while it is only about 1,400 meters in the mountains of North America. In Taiwan, this line occurs around 3,600 meters, and in the South-east Asia, it can occur from 4,500 to 5,000 meters and up.

Fir（*Abies kawakamii* (Hayata) Ito）sticks close to the forest line, so the fir is the highest-altitude forest. *Abies kawakamii* grow trunks straight, reaching heights of 20 to 30 meters. The form of these trees is very similar to a Christmas tree, and this is the forest that is also called the Black Forest. Just as in tundra regions, the withdrawal of the glaciers northward meant that plant species have retreated to higher latitudes and higher altitudes, so the distribution of subalpine conifer forest in Taiwan is between 3,000 and 3,500 meters, and these are closely linked to conifer forests south of the Arctic region. Conifer forests in Arctic, Siberia and North America have similar tree shapes and forest structure -- straight main trunks without forks and a single-layer canopy.

Hemlock forests are distributed at lower altitudes than the *Abies kamakawii*. *Tsuga chinensis* (Franch.) Pritz. ex Diels var. *formosana* (Hayata) Li & Keng are not straight like *Abies kamakawii*, but are more like broadleaf trees in their bending. Taiwan's hemlock forests are generally found between 2,500 and 3,000 meters in elevation, and of all the hemlock forests worldwide, only those in Taiwan and Sichuan provide habitat for the panda. The hemlock forests, like those of *Abies kamakawii*, have only two levels, a canopy of hemlock and shrub layers of *Yushania niitakayamensis* (Hayata) Keng f.. At the same altitudes with the hemlock are *Picea morrisonicola*

瑞士阿爾卑斯山，森林的分布只能到達海拔2,000公尺左右，在北美洲北部山地則只能發育到1,400公尺左右，而在台灣則可達海拔3,600公尺，在南洋地區可能要在4,500~5,000公尺以上。

緊接於「森林界線」下的是「冷杉林」，所以冷杉林是海拔最高的森林。冷杉（*Abies kawakamii* (Hayata) Ito）樹幹筆直，樹高可達20~30公尺，與國外的聖誕樹造型相仿，也就是所謂的黑森林。和寒原地區一樣，由於冰河北退的關係，生物種類分別退到高緯度及高海拔地區，因此分布在台灣海拔3,000~3,500公尺的亞高山針葉林帶，與北極圈以南的針葉林帶具有密切的關係。在北歐、西伯利亞、北美洲的針葉林，樹型、森林結構與台灣的冷杉林相似—主幹筆直不分叉，喬木層僅有一個層次。

鐵杉林是海拔分布較冷杉低的針葉森林，鐵杉（*Tsuga chinensis* (Franch.) Pritz. ex Diels var. *formosana* (Hayata) Li & Keng）樹幹不像冷杉筆直，反而像闊葉樹般彎曲。台灣的鐵杉林大約在海拔2,500~3,000公尺之間，鐵杉林在全世界的分布只在台灣和中國四川一帶，就是熊貓生活的舞臺。鐵杉林和冷杉林一樣，祇有兩層，冠層是鐵杉，地被層則是玉山箭竹。與鐵杉在同一海拔高度的還有雲杉（*Picea morrisonicola* Hayata）林，在氣候較潮濕的地區，如思源埡口至南湖北山一帶，則是雲杉林而非鐵杉林。

在海拔2,500~3,500公尺一帶，有時可見一株株樹幹白色，全無葉子的「樹木」矗立箭竹草原間，這樣的樹林稱為「白木林」。白木林其實是冷杉或鐵杉林火災之後殘留下來的樹幹，因為長期受到風吹雨淋，焦黑的外皮已經消失，剩下裡面的木材，然而因為高海拔地區枯木腐化的速度較慢，所以不容易腐爛分解，就會看到在低海拔地區看不到的「白木林」景觀。白木林中樹幹較筆直的是冷杉，而樹幹分枝較多的則為鐵杉。

台灣的冷杉林和鐵杉林受到干擾之後會變成箭竹原，因為地被層的箭竹具有深埋地下的地下莖，經大火之後仍然倖存，且能快速抽出新芽，所以在冷杉林或鐵杉林森林大火之後，會呈現一片箭竹原，所以玉山箭竹（*Yushania niitakayamensis* (Hayata) Keng f.）原的出現，即代表原來生

Hayata forests, found in places with a somewhat wetter climate. such as from the Ssuyuan to Mt. Nan-Hu-Pei-Shan, where fir is seen instead of hemlock.

In the areas between 2,500 and 3,500 meters, at times individual trees with white trunks can be seen. These totally leafless "trees" stand among Sinarundinaria nitida, and this type of tree is called a "white wood". These "white woodland" are actually the leftover trunks from hemlock or Abies forests that have been destroyed by fire, which have been acted on by wind and rain for a long period of time, stripping away their dark colored outer bark and exposing the wood inside. Afterwards, because of the high altitude, wood rot progresses slowly, and the trees do not decompose easily. This is why these "white forests" can be seen at such altitudes. In the white forests, the straight trunks are Abies, while the branching trunks are mostly hemlock.

After being disturbed, Taiwan's fir or hemlock forests may become areas of *Yushania niitakayamensis* (Hayata) Keng f.. This is because the ground cover layer of *Yushania* have deep roots and they are able to live through a fire. They quickly send up tall new growth, so that after a hemlock or Abies forest is burned, *Yushania* appear. The appearance of *Yushania niitakayamensis* means that the original forest has already been destroyed by fire. That is to say that the *Yushania niitakayamensis* at high-elevations in Taiwan shows that an original stand of hemlock or fir has been destroyed. After a fire, the fir forest will first come back with *Yushania niitakayamensis*, and, if seeds are available, the fir will gradually come back afterwards. Hemlock forest also shows *Yushania niitakayamensis* growth first after a fire, and then this is replaced by *Pinus taiwanensis* Hayata, with the hemlock restoring itself afterwards. The distribution of *Yushania niitakayamensis* supercedes the subalpine conifer forest belt and the arctic conifer belts, for which reason it can be seen that the distribution of pioneer plants is

■冷杉的毬果。 攝影／蕨類研究室

■369山莊的白木林。 攝影／蕨類研究室

broader, and that of successor plants somewhat narrower.

The high altitude regions are a major source of Taiwan's rivers, and currently they are mostly protected in national parks. However, many people cause forest fires out of carelessness, which has converted many of these forests into groves of Yushania niitakayamensis. This has a large impact on the nutrient content of the water.

長該地的森林已經被大火燒毀，也就是說，台灣高海拔出現大面積的箭竹原，顯示原有的冷杉林或鐵杉林已泰半消失。火災之後的冷杉林，會先形成箭竹原，在冷杉種源不匱乏的情況下會逐漸恢復成冷杉林。而火災之後的鐵杉林在形成箭竹原之後，會先演替成台灣二葉松（Pinus taiwanensis Hayata），再恢復成鐵杉。玉山箭竹原的分布可說是跨越了亞高山針葉林帶與冷溫區針葉林帶，由此可看出屬於先鋒型的植物群落分布範圍較廣，而演替上較晚出現的植物群落分布範圍較窄。

高海拔地區是台灣主要大河的發源地，目前多受國家公園保護，然而許多人為疏失造成森林大火，使得很多地方都成為箭竹草源，對於水份的涵養有極大的影響。

■雲杉生長在比較潮溼的地區，分布海拔高度與鐵杉相同。 攝影／蕨類研究室

漫步在雲端

鐵杉－雲杉林

Hemlock-Spruce Forest

◎撰文／劉靜榆、曾彥學
◎ Text ／ Ching-Yu Liou
Yen-Hsueh Tseng

台灣鐵杉在海拔2,500公尺以上形成優勢森林，另一樹種台灣雲杉則在同一海拔地帶之不同生育地形成優勢林型，構成所謂「鐵杉－雲杉林帶」。一般而言，台灣鐵杉之樹冠成傘型，樹幹的分枝多，樹型似闊葉樹，多生長於向陽坡或山脊較爲乾旱之處。台灣雲杉之樹冠呈三角型，主幹通直圓滿，高可參天，喜好背陽之坡面且土壤層深厚之立地。台灣鐵杉因爲材無大用，而免於被大量伐採的命運，成爲台灣木材蓄積量最高的樹種。台灣雲杉的族群數量雖不及鐵杉，但卻是全世界各種雲杉中，唯一向南跨越北回歸線的樹種，在植物地理學上有其特殊意義，而且台灣雲杉還是台灣的特有種。就更新機制而言，台灣鐵杉可以在老熟林內，利用孔隙內新生之幼苗更新，台灣雲杉之幼苗則無法在鬱閉林分下更新，其更新孔隙爲林內干擾發生之地點，如溪源及山坡上側之崩塌與地滑等。

Taiwan's hemlocks are predominant in the forests from 2,500 meters of altitude and up. Another species, the Taiwan spruce, is predominant at the same altitude, but in different types of areas. These two species form the "Hemlock-Spruce Forest Zone." Generally speaking, Taiwan hemlock has an umbrella-shaped canopy, and a trunk with multiple branches. The shape of these trees is like that of broadleaf trees, and most grow on shady slopes or on dry areas atop mountain ridges. The canopy of Taiwan spruce is triangular in shape, with straight, round main trunks reaching up to the sky. This species prefers backlit areas with relatively deeper soil. Taiwan hemlock is not commonly used for lumber, so it has avoided being overharvested, and has become the main species in Taiwan's lumber reserve. Although there are fewer Taiwan spruce than hemlock, of all the spruces of the world, this is the only species that has ventured the Tropic of Cancer, and it has special significance in plant geography. Moreover, the Taiwan spruce is a species unique to Taiwan. In terms of renewal mechanisms, in mature forests the Taiwan hemlock is capable of using new seedlings in tiny openings to produce new growth, while the young seedlings of the Taiwan spruce cannot renew themselves in close-stand forest; it finds its niche in disturbed areas of the forest, such as areas of landslides.

■鐵杉未成熟之毬果。

攝影／賴國祥

台灣木材蓄積量最高的樹種

鐵杉屬（*Tsuga*）隸屬松科（Pinaceae），全世界共有16種，產於東亞及北美，其分布及演化雖未至孑遺狀態，然已呈不連續分布，目前僅存於北半球中緯度較為溫暖潮濕之生育地。

台灣的鐵杉僅有一種一台灣鐵杉（*Tsuga chinensis* (Franchet) Pritz. *ex* Diels var. *formosana* (Hayata) Li & Keng），為台灣特有變種，分布於中央山脈海拔 1,800~3,200 公尺之地區，於海拔 2,500~3,000 公尺間多為純林，本林型之水平分布北起拉拉山、塔曼山及巴博庫魯山，南至大武山。垂直分布之海拔高度下限，約與台灣山地盛行雲霧帶之上側相當。由於各地雲霧帶海拔高度略有不同，鐵杉林之分布海拔亦隨著變化。台灣中部的鐵杉純林大約分布在2,500公尺以上，混合林約在海拔1,700公尺處便可發現，台灣北部之拉拉山及塔曼山則分布在1,900公尺以上。

台灣鐵杉之最適生育地依台灣山地植群帶之區分屬山地上層針葉林帶，其氣候屬涼溫帶（cool-temperate），年均溫度8~11℃；溫量指數（warmth index）為36~72℃，計算溫量指數的基礎是由於林木生長之開始溫度為5℃，因此將一年之中，月平均溫超過5℃者，各減去5後，再累加起來。台灣鐵杉於海拔3,000公尺以上多與台灣冷杉（*Abies kawakamii* (Hay.) Ito）混生，海拔較低處則與台灣雲杉（*Picea morrisonicola* Hay.）、華山松（*Pinus armadii* Franch.）、台灣二葉松（*Pinus taiwanensis* Hay.）、台灣扁柏（*Chamaecyparis obtusa* Seib. & Zucc. var. *formosana* (Hay.) Rehder）、紅檜（*Chamaecyparis formosensis* Matsum.）、台灣杉（*Taiwania cryptomerioides* Hay.）、香杉（*Cunninghamia konishii* Hay.）等針葉樹混生，其他闊葉樹為其下層林冠。

台灣鐵杉為台灣木材蓄積量最高的樹種，其木材可作木漿、建築及家具等，屬於針二級木，其樹冠多呈傘型，主幹的分枝多，樹型似闊葉樹。依據柳榗等人於1961年調查台灣主要林型時，由全島11個鐵杉林樣區的統計資料中，鐵杉之平均材積為1,200 m^3／ha，平均樹高24公尺，平均胸高直徑62公分，最高優勢木年齡約650年。台灣現存之鐵杉林之分布由南到北，範圍很廣，海拔差異達1,400公尺，其型態特徵、生育地環境、更新情形及伴生植物種類略有不同，在插天山自然保留區調查報告中，拉拉山（2,031公尺）西側稜線上之鐵杉純林，其胸高直徑達60公分。而中部的沙里仙溪上游之鐵杉林最大者胸徑為280公分。另按蘇鴻傑教授於1992年在北大武山調查，36個樣區中鐵杉之胸高斷面積佔87%，且各種直徑充分出現，最大者胸徑為302公分，樹高42公尺。其相關的文獻大部分都是植群調查時對範圍內之鐵杉林加以描述，以中部山區沙里仙溪集水區為例，鐵杉林型主要分布於海拔2,300~3,200公尺，坡度30~65°，含石率1~5級，多位於東北及北向坡面，全天光與直射光空域皆幾達百分之百，林分組成上層為台灣鐵杉，下層為玉山箭竹（*Yushania niitakayamensis* (Hay.) Keng. f.），林內竹桿高大。其他闊葉樹有厚葉柃木（*Eurya glaberrima* Hay.）、台灣鴨腳木（*Schefflera taiwaniana* (Nakai) Kanehira）、玉山假沙梨（*Stranvaesia niitakayamensis* (Hay.) Hay.）、台灣杜鵑（*Rhododendron formosanum* Hamsl.）、玉山杜鵑（R. *pseudochrysanthum* Hay.）、台灣馬醉木（*Pieris taiwanensis* Hay.）、玉山灰木（*Symplocos anomala* Brand）、小實女貞（*Ligustrum microcarpus* Kanehira & Sasaki）及川上氏小蘗（*Berberis kawakanii* Mizush)等，其中台灣鴨腳木之小苗常成群密生。地被植物多為蕨類，如：台灣瘤足蕨（*Plagiogyria glauca* (Blume) Merr. var.

■鐵杉主幹分叉多，側枝較粗大，樹冠寬廣。 攝影／賴國祥

■台灣鐵杉。 攝影／賴國祥

philippinensis Christ）、柄囊蕨（*Peranema cyatheoides* Don.）、阿里山鱗毛蕨（*Dryopteris squamiseta* (Hook.) Ktze.）、小膜蓋蕨（*Araiostegia perdurans* (Christ) Copel.）等。

由於鐵杉的生育地常是陡峻難以到達，其木材堅硬而且易龜裂，因此只有部份林分曾被伐採。柳楷等人認為台灣的鐵杉會成為純林的原因應是氣候及立地之因素，由林下幼樹種類及數量來看，應屬安定集團植生。按蘇氏於北大武山台灣鐵杉之族群動態研究中顯示，在鐵杉林帶中，台灣鐵杉林之發育階段有三種塊集，分別代表孔隙期、建造期及成熟期，這些塊集普遍鑲嵌於鐵杉林中，其分布與林分之環境因子沒有相關。該研究指出，台灣鐵杉的族群構造多呈安定的反J形分布，其幼苗可以在鬱閉林分下更新，甚至在老木單株枯死之小裂隙下（100~400平方公尺）幼苗即可能出現。

台灣高山森林易受各種自然干擾(如火災、颱風及山崩等)影響，例如在麟芷山附近之鐵杉林，曾於1963年5月發生火災而形成大面積之箭竹草原，其箭竹高約50公分，生長極為濃密，混生其間的植物有玉山假沙梨（*Stranvaesia niitakayamensis* (Hay.) Hay.）、台灣紅榨槭（*Acer morrisonense* Hay.）、高山白珠樹（*Gaultheria itoana* Blume）、玉山金絲桃（*Hypericum nagasawai* Hay.）、玉山石松（*Lycopodium vditchii* christ）、玉柏（*Lycopodiu juniperoideum* Sw.）、台灣藜蘆（*Veratrum formosanum* Loesen. f.）、鼠麴草（*Gnaphalium affine* D. Don）、高山柳葉菜（*Epilobium amurense* Hausskn.）、玉山小米草（*Euphrasiatran morrisonensis* Hay.）及龍膽類（*Gentiana* spp.）之植物，次生之鐵杉稚樹數量極多，胸徑約20公分，鐵杉林與玉山箭竹間已形成一明顯之推移帶。在干擾較少的地區，鐵杉胸徑常高達200公分以上，老齡的鐵杉林冠層寬廣而重疊，森林底層很少有直射光，且林下密生的箭竹阻礙鐵杉幼苗更新，但若林內有大型風倒木致使周圍礦質土裸露，加以冠層疏開，林床光照度增加，其下層鐵杉及其他闊葉樹之幼苗皆得以生長。國外研究孔隙動態之學者發現，有多種鐵杉能在孔隙形成前即已出現更新者，當孔隙發生後，苗木迅速生長，而達到林冠層。因此若按徑級推測，老齡的鐵杉林多呈異齡結構，利用孔隙輪番更新，可使林分持續不衰。

台灣鐵杉在海拔2,500公尺以上形成優勢森林，另一樹種台灣雲杉則在同一海拔地帶之不同生育地成優勢林型。一般而言，台灣鐵杉多生於向陽坡或山脊較為乾旱之處，而台灣雲杉則喜好背陽之坡面且土壤層深厚之立地。

一柱擎天

雲杉屬（*Picea*）隸屬於松科，全屬約40種，均產於北半球寒帶，台灣雲杉(*Picea morrisonicola* Hay.)是唯一向南跨越北回歸線的種，綜觀整個雲杉屬之地理分布，台灣為其分布下限，就植物地理學而言有其特殊意義。台灣雲杉為台灣特有種，其樹冠成三角型，主幹通直圓滿，高可參天，分布於中央山脈海拔2,100~3,200公尺之地區，其中僅沙里仙溪、楠梓仙溪上游及南湖大山雲稜山莊形成純林外，其它多散生於台灣冷杉及台灣鐵杉林中。本林型之分布北起南湖大山，沿中央山脈山脊兩側地帶南行，止於關山，常生長於山腰或山麓，鮮見純林，與之混生的樹種除台灣鐵杉外，尚有華山松、台灣冷杉、台灣扁柏、紅檜、香杉、威氏粗榧（*Cephalotaxus wilsoniana* Hay.）。

由於台灣雲杉喜好背陽之陰坡，土層肥沃、深厚之立地，故林下植生豐富，競爭亦強烈，上層樹冠略有疏開則其下層即被其他闊葉樹侵入，或為頂芽狗脊蕨（*Woodwardia unigemmata* (Makino) Nakai）等密覆，表土無法裸露。以中部山區沙里仙溪集水區為例，台灣雲杉林型主要分布於玉山前峰向北北西坡面海拔約2,500~3,000公尺之間，多為純林。台灣雲杉之分布下限可延伸至海拔2,100公尺左右，其生育地之坡度約28~42度，向北或西北方，土壤層深厚，含石率1~2級，全天光空域45~59%。直射光空域39~56%。上層樹冠為台灣雲杉優勢，混生有華山松、台灣紅檜、台灣鐵杉或紅豆杉（*Taxus mairei* (Lemee & L'evl.) Hu *ex* Liu）等針葉樹，下層灌木以闊葉樹為主，如玉山木薑子（*Litsea morrisonensis* Hay.）、漸尖葉新木薑子（*Neolitsea acuminatissima* (Hay.) Kanehira & Sasaki）、小實女貞（*Liqustrum microcarpum* Kanehira & Sasaki）、太平山莢迷（*Viburnum foetidum* Wall.

■鐵杉雄毬花。 攝影／賴國祥

var. *rectangulatum* (Graeb.) Rehder）、厚葉柃木、刺格（*Osmanthus heterophyllus* (Don) Green var. *bibracteatus* Hay. Green）、阿里山十大功勞（*Mahonia oiwakensis* Hay.）、刺果衛矛（*Euonymus echinatus* Wall.）等。雲杉林分布至海拔較低處則混入大量闊葉樹種，形成第二層闊葉樹冠層，其組成主要有昆欄樹（*Trochodendron aralioides* S. et Z.）、台灣紅榨槭、薄葉虎皮楠（*Daphniphyllum himalaense* (Benth.)Muell.-Arg. *ssp. macropodum* (Miq.) Huang）、高山鴨腳木及狹葉櫟（*Cyclobalanopsis stenophylloides* (Hay.) Liao）等。本林型之地被植物在海拔較高處常見有冷杉異燕麥（*Helictotrichon abietetorum* (Ohwi) ohwi）、玉山鹿蹄草（*Pyrola morrisonensis* (Hay.) Hay.）、阿里山假寶鐸花（*Disporum arisanensis* Hay.）、頂芽狗脊蕨等；在海拔較低處以曲莖蘭嵌馬藍（*Parachampionella flexicaulis* (Hay.) Hsieh & Huang）、松田氏冷水麻（*Pilea matsudai* Yamamoto）等較為優勢。

■台灣雲杉樹形高大，可達45公尺。

攝影／賴國祥

台灣雲杉因受人為伐採及各種自然干擾影響，以致僅存之雲杉林往往殘破不堪，又由於本種並無極高之經濟價值，因而甚少造林，且根據文獻及調查紀錄，其幼苗甚為罕見。柳榗將台灣的雲杉林依其樹種組成變化推論：台灣雲杉為演進過渡時期之森林，屬不安定集團植生。而按蘇鴻傑教授於大甲溪上游取樣調查結果，當先驅之台灣二葉松僅存老樹時，台灣雲杉仍有少量苗木，故推測台灣雲杉為出現於陽性樹種之後的中性樹。而由沙里仙溪集水區內台灣雲杉林之族群動態研究中顯示，台灣雲杉之幼苗無法在鬱閉林分下更新，甚至在老木單株枯死之小裂隙下，幼苗亦不出現，台灣雲杉之更新孔隙為林內干擾發生之地點，如溪源及山坡上側之崩塌與地滑等。台灣雲杉更新所需的面積遠比單株死亡所產生之冠層孔隙為大，由於在此種孔隙內之某一時段大量出現幼苗，故在其後繼續發育之建造期林分中，其齡級分布多呈同齡林之鐘形分布。台灣雲杉林在不同地點之極盛相性質，如忽略生育地條件，可能導致不同之推測，究竟台灣雲杉是次生林演替之中途階段，抑或為具有塊集構造之極盛相林型？

國外研究孔隙動態之學者發現鐵杉能在孔隙形成前即已出現更新者，雖然國內的學者則認為，台灣鐵杉之更新材料是孔隙新生之幼苗，而非前生苗，但鐵杉可以在老熟林內更新，而成為蓄積量最高的樹種。反觀台灣雲杉之更新孔隙為林內干擾發生之地點，其幼苗無法在鬱閉林分下更新。不同的更新機制，會形成不同的塊集，按塊集動態之理論，極盛相森林由不同發育階段之塊集組合而成，塊集之間會有動態變化，並呈平衡狀態存在，因此無論其更新機制為何，若可避免人為強力的干擾，則鐵杉一雲杉林帶應可永續存在。

■台灣雲杉的球果。 攝影／賴國祥

■鐵杉與草生地之推移帶。攝影／賴國祥

代表性伴生植物

厚葉柃木

Eurya glaberrima Hay.

山茶科 Theaceae

台灣特有種，產於海拔2,000~3,000公尺的山區，它是一種常綠的小喬木，但常見到的都是幼樹，故呈灌木狀。在分類上，屬於山茶科（Theaceae）的柃木屬（*Eurya*）植物，顧名思義厚葉柃木之葉片較其他同類植物厚，在台灣原生柃木類植物中，它的分布海拔最高。

攝影／賴國祥

威氏粗榧

Cephalotaxus wilsoniana Hay.

三尖杉科 Cephalotaxaceae

本種為三尖杉科之植物，葉線狀鐮刀形，氣孔帶白色，種實成熟時紫色。為台灣特有植物，分布於台灣山區海拔1,400~2,400公尺之針闊混交或針葉樹林中，為下層植物，多為單株散生，未見其集生呈群落，生長在向陽之岩質陡坡上或茂密之森林中，種子發芽困難，可見於雲杉林之下層。

攝影／曾彥學

玉山箭竹

Yushania niitakayamensis (Hay.) Keng. f.

禾本科 Gramineae, Poaceae

分布於台灣與菲律賓，是台灣高海拔山區的代表性植物之一。玉山箭竹為台灣分布海拔最高的竹類，常在陽光極強的南向坡形成大片「草原」。玉山箭竹是一種適應力極強能屈能伸的植物，在強風盛行的環境上，株高不及30公分，但在有森林的庇護環境下可高達4公尺。

攝影／賴國祥

玉山木薑子

Litsea morrisoensis Hay.

樟科 Lauraceae

為台灣特有種，產於海拔2,000~3,000公尺的山區。屬於樟科植物成員之一，是台灣原生樟科植物中分布海拔第二高的樹種，僅次於漸尖葉新木薑子。它與漸尖葉新木薑子最大的區別，在於它的葉子互生，而後者則是叢生在枝條先端。

攝影／曾彥學

台灣鴨腳木

Schefflera taiwaniana (Nakai) Kanehira

五加科 Araliaceae

為台灣特有種，產於海拔2,000~2,500公尺的山區。屬於常綠的小喬木，小葉片常5~8片集生在一齊，外觀似鴨掌狀，故稱台灣鴨腳木。在分類上，屬於五加科（Araliaceae）的鴨腳木屬（*Schefflera*）植物，與台灣早期做火柴棒的材料江某（*S. octophylla*）屬於同一家族。

攝影／曾彥學

玉山假沙梨

Stranvaesia niitakayamaensis (Hay.) Hay.

薔薇科 Rosaceae

為台灣特有種，產於海拔2,000~3,000公尺的山區，屬於薔薇科（Rosaceae）的一種小喬木，每當晚秋時節，橙紅的果實佈滿樹梢，在落葉飄零的季節裡，格外鮮艷奪目，是鐵杉林帶最具代表性植物之一。

攝影／賴國祥

台灣馬醉木

Pieris taiwanensis Hay.

杜鵑花科 Ericaceae

分布於中國大陸華南及台灣，台灣主要產於中部2,000~2,700公尺的山區，但北部可降至陽明山500公尺的山區。屬於杜鵑花科（Ericaceae）家族的成員，植株具有麻醉效用，傳聞馬吃了之後，站立不穩，如醉酒一般，故名馬醉木。

漸尖葉新木薑子

Neolitsea acuminatissima (Hay.) Kanehira & Sasaki

樟科 Lauraceae

為台灣特有種，是台灣原生樟科（Lauraceae）植物中分布海拔最高的成員，可達3,100公尺，故又名高山新木薑子。常綠小喬木，在鐵杉林中屬於第二層冠木層的優勢樹木，因此很容易見到其蹤跡。

刺格 *Osmanthus heterophyllus* (Don) Green var. *bibracteatus* (Hay.) Green

木犀科 Oleaceae

分布於中國大陸及台灣，台灣產於高海拔山區。本種與常見的桂花（O. fragrans）同屬於木犀科（Oleaceae）植物家族，它的葉子變化很大，有時全緣，有時疏鋸齒緣，不太容易辨識，也因為上述特徵，又名異型葉木犀。

攝影／曾彥學

太平山莢迷 *Viburnum foetidum* Wall. var. *rectangulatum* (Graeb.) Rehder

忍冬科 Caprifoliaceae

屬於忍冬科（Caprifoliaceae）植物成員之一，分布於中國大陸及台灣，在台灣山區分布的海拔範圍很廣，從1,600~3,300公尺都有它的蹤跡，為鐵杉林下層植被常見的灌木。

攝影／曾彥學

玉山鹿蹄草 *Pyrolla morrisonensis* (Hay.) Hay.

鹿蹄草科 Pyrolaceae

為台灣特有種，產於海拔2,300~3,200公尺的山區。屬於鹿蹄草科（Pyrolaceae）植物的成員，它是一種小草本，常見於鐵杉林底層的地被，每年6~7月間可看到它由根莖抽長的花莖懸著數朵白色的小花。

攝影／賴國祥

墨綠色的畫布

台灣冷杉林

The Taiwan Fir Forest

◎撰文／賴國祥

◎Text ／Kwo-Shang Lai

台灣3,000公尺以上的地區，常見青黃色的草生地與墨綠色的森林交織成一片高山特有的景觀。這墨綠色的森林絕大部份就是台灣冷杉林。記得大二時，第一次上台灣的高山，初春時節走在台灣冷杉林下，猶見積雪處處，那種感覺與印象讓我久久無法忘懷，從此愛上台灣的高山，愛上台灣冷杉。台灣冷杉廣泛分布於台灣3,000公尺以上的亞高山地區，4~5月間或可見暗紅而略帶紫氣的雌毬花直立於枝梢，到了10月，成熟的毬果更展現一種獨特的藍紫黑色，伴隨雪白色之樹脂，色彩艷麗而迷人，相當的特別、耀眼。台灣冷杉樹幹通直、圓滿，側枝平展而略爲下垂，樹高可達35公尺，樹冠整體而言爲圓錐形，其形成之純林亭亭聳立，爲台灣亞高山地區特殊美景之一，漫步於林下，保證令人心曠神怡。台灣冷杉爲冷杉屬植物於地球上分布之南界，台灣的高山因不易到達，目前比較沒有開發的壓力，應該不會有消失的危機，但它是一種具有無限潛力的生態景觀資源，宜善加保育，永續經營。

The areas around 3,000 meters in elevation in Taiwan often boast grassy areas of yellow and green and dark-green forests interwoven into a landscape unique to these mountains. The vast majority of the trees in the dark green of the forest are *Abies kawakamii* (Hayata) Ito. Once, when I was a sophomore in college, the first time I visited the high mountains of Taiwan, I walked among the Abies forest in early spring where occasional patches of snow could still be seen. This feeling and the impact it created are things I will never forget. From that moment, I fell in love with Taiwan's high mountains, and with the *Abies kawakamii*.

■台灣冷杉之雌毬花，直立於枝梢，暗紅略有紫氣。

攝影／賴國祥

分類地位、形態及分布

台灣冷杉（*Abies kawakamii*(Hayata) Ito）為台灣特有種，廣泛分布於台灣亞高山地區，最早的採集記錄是日本人本多靜六於1896年前往玉山所採獲。台灣冷杉與香青（玉山圓柏）（*Juniperus squamata* Buch.-Ham.）為台灣寒帶林或稱亞高山針葉樹林（subalpine coniferous forest）之主要組成樹種，在此林帶中尚有少數溫帶林或稱冷溫帶山地針葉樹林（cold-temperate montane coniferous forest）向上延伸的台灣鐵杉（*Tsuga chinensis* (Franch.) Pritz. ex Diels var. *formosana* (Hayata) Li & Keng）、台灣雲杉（*Picea morrisonicola* Hayata）、台灣華山松（*Pinus armandii* Franchet var. *masteriana* Hayata）、台灣二葉松（*Pinus taiwanensis* Hayata）等。台灣3,000公尺以上的高山，由北至南除北大武山外，都可以看得到台灣冷杉，為台灣海拔分布僅次於香青的針葉樹種。台灣冷杉屬裸子植物門，松柏綱，松柏部，松科（Pinaceae），冷杉屬之植物，其樹皮灰褐色，葉闊線狀，扁平，長1~1.5公分，基部狹，先端稍廣，有凹缺或圓形鈍頭。雄毬花黃色略紅，長於枝條先端的葉子下方。雌毬花直立，暗紅略有紫氣。毬果成熟時為藍紫黑色，橢圓形，先端常滲泌出雪白色之樹脂，直立於小枝上，自綠色針葉間聳出，極為耀眼。成熟後，果鱗（cone scale）及種子皆離中軸片片落下，種子帶有頂翅，可隨風飄散。台灣冷杉樹高可達35公尺，

■台灣冷杉：主幹通直圓滿不分叉，側枝細而平伸且略下垂，樹冠整體而言為圓錐形。 攝影／賴國祥

然一般為25公尺左右。樹齡則大多在500年以下，超過500年者較為罕見。

台灣冷杉的分布中心介於海拔3,000~3,500公尺，常形成大面積純林，由於台灣冷杉樹形峭聳、樹姿端麗，形成之純林亭亭聳立，為台灣亞高山地區特殊美景之一，其上至3,600公尺常與香青混生，而下至3,100公尺則與鐵杉混生，形成過渡地帶。

■台灣3000公尺以上之地區，森林與草生地呈現一種鑲嵌式（mosaic）之植被景觀。左下方為森林與草生地間之推移帶。 攝影／賴國祥

台灣的高山環境

台灣的高山為台灣冷杉的主要分布地帶，這個區域由於雨水多，河川侵蝕旺盛，加上地盤間歇性上昇，使溪流不斷回春，下切加強，導致台灣之高山不但絕對高度甚大，且相對地勢極陡，台灣冷杉林內一般地勢崎嶇，落差極大。

台灣超過3,000公尺的山峰有230幾座，但高山上每可發現早期地形侵蝕殘留之台地及台地上之湖泊，此等湖泊面積雖小，但湖水清澈，加上四周森林或草原坡的相互輝映，景緻秀麗動人。至於聯絡各高峰間之山脊，多數亦頗平坦而成為登山或連峰縱走較易之路線。

■高山湖泊：雪山翠池，為台灣海拔最高之高山湖泊，其旁為香青林。

攝影／賴國祥

台灣高山之氣候為寒帶重濕氣候，冷而多濕，冬寒有積雪，一般12月底開始下雪，至隔年3月為雪季。平均溫度以玉山及合歡山為例，1月為-2.3℃與2℃，7月為2.7℃與10℃。年平均雨量約3,500公釐，年平均降水日數145天，主要集中於5~6月，12~1月最少。平均相對溼度約80%，霜期則達6個月。由於雨量豐沛，限制植物生長的因子主要為溫度。

本區地質大部分是由堅硬或是經過輕度變質的第三紀巨厚沈積岩組成，區內大部分的沈積物是深灰或灰黑色劈理良好的硬頁岩（argillite）、板岩（slate）以及千枚岩（phyllite），也就是經過變堅或變質的泥質岩石。這些岩石中常含有許多白色的小石英脈。至於土壤以石質土為主，灰壤或棕色灰化土則存於地形較緩、排水良好之處。至於針葉樹林、草生地及其間推移帶之土壤質地大都為壤土、坋質壤土或黏質壤土，土色則屬黑色至棕色系列。

■台灣高山冬天常會下雪，此為台灣冷杉為雪所覆蓋之景況。

攝影／賴國祥

物候（phenology）特性與苗木建立（establishment）

台灣冷杉4~5月開花，海拔較低者開花較早，與較高海拔地區相差可達一個月。10 月底種子開始飄落，部份種子於冬季為雪所覆蓋，翌年5~7月發芽，7~8 月有大量幼苗發生。

台灣冷杉之毬果於 10 月底成熟後，種子開始散佈，12月及次年1月為種子散佈之高峰期，而後漸減，至4 月後銳減，7 、8 、9 月已幾無種子飄落，若有，也僅為殘留樹上受強風吹落者。台灣冷杉一年之總下種量，可高達每公頃126萬粒，然亦有低至每公頃0.5 萬粒者，可見其豐、歉年極為明顯，一般台灣冷杉 4~5 年會有一次結實豐年。

台灣冷杉屬偏陽性樹種或孔隙更新樹種（gap-phase species），除非原有冷杉老死，或受其他因子干擾（如火燒、颱風）而形成無林木地或倒木孔隙，否則無法更新。其苗木之建立以森林邊緣為主，林內僅於倒木孔隙才會有苗木發生，因為林內光度不足，除非遇有大孔隙，否則種子發芽後，無法存活。而林緣之苗木又因種子散布距離之關係，大都集中於距林緣10~20公尺之範圍內，與玉山箭竹（*Yushania niitakayamensis* (Hayata) Keng f.）草生地間形成一明顯的推移帶（ecotone）。

上■台灣冷杉之雄毬花，黃色略紅，長於枝條先端的葉子下方。

下■台灣冷杉林：台灣冷杉樹形峭聳、樹姿端麗，形成之純林亭亭聳立，為台灣亞高山地區特殊美景之一。 攝影／賴國祥

■台灣冷杉成熟之毬果：藍紫黑色，橢圓形，先端常滲泌出雪白色之樹脂。 攝影／賴國祥

推移帶之火燒週期

台灣3,000公尺以上之地區，森林與草生地呈現一種鑲嵌式（mosaic）之植被景觀。南向坡山腹至陵線較乾燥、平緩之處，常可見大面積之草生地，森林則分布於北向坡及南向坡的山腹至山谷較陰濕之處，這主要的原因就是火燒（fire）所造成。根據調查台灣冷杉與草生地間推移帶之火燒週期（fire period）約為80年，也就是說平均每隔80年會發生一次火燒。但因為濕氣等因素之影響，一般火燒至林緣即會熄滅，僅林椽外小苗被火燒死，林內母樹不會受到影響。但若發生強烈火燒，由於台灣冷杉樹葉富含油脂，燃燒迅速，而樹幹含水量高不易燃，火後常形成所謂「白木林」。

■ 雪山白木林：為台灣冷杉林經火燒後所形成。　攝影／賴國祥

動態平衡

台灣冷杉林入侵草生地之間隔，因受結實年齡（冷杉類約需30~50年生才會結實，產生有效種子）或立地環境之關係（其必需形成林緣效應(edge effect)如增加蔽蔭、濕氣後，才有利苗木建立），平均長達36年。其入侵方式為一次長距離，在此距離內建立多數苗木後再一次長距離入侵，而非漸進式，也不是一次建立，可說是逐年階段性入侵。可知，若無人為影響，無火燒，草生地將被侷限於只適合它們的地形與土壤之環境，而其他之林木將佔有較廣大之地區。依苗木之入侵間隔及距離，每百年約有三次之苗木建立，入侵距離或說推移帶寬度可達30公尺。但在此段時間，入侵之苗木又常因火燒而死亡，致使其入侵常因火燒之干擾而退回。即台灣冷杉以階段性的方式入侵草生地，然因週期性火燒之關係而退回原處，其間保持一動態平衡之關係。可知在台灣的亞高山地區，只要干擾體制，如火燒，仍然存在，其森林與草生地獨特的鑲嵌式植被景觀亦將永遠存在。

■ 海拔3,000~3,100公尺左右之台灣冷杉與鐵杉過渡帶，其中樹冠寬廣，側枝較粗大者為鐵杉。　攝影／賴國祥

■ 台灣冷杉林：台灣冷杉樹形峭聳、樹姿端麗，形成之純林亭亭聳立，為台灣亞高山地區特殊美景之一。　攝影／賴國祥

主要伴生物種

香青（玉山圓柏）

Juniperus squamata Buch.-Ham.

柏科 Cupressaceae

台灣海拔分布最高的樹木，其形態可分喬木及灌木兩種。葉披針形，顏色淡墨綠，先端銳尖，稍刺人。4~5月開花，12月後果熟，毬果球形或橢圓形，熟時黑紫色。因受山地氣候影響，灌木型的香青形狀變化萬千，普遍見於台灣3,000公尺以上之向陽開闊地或山脊稜線上，並常與玉山杜鵑群集叢生，形成高山矮盤灌叢景觀。喬木狀的香青林大多發現於坡度平緩，土壤發育較佳的避風谷地，以雪山、南湖大山、玉山及秀姑巒山較具規模。

狹葉七葉一枝花

Paris lanceolata Hayata

百合科 Liliaceae

多年生草本，莖單出直立，葉狹披針形，全緣，7~11枚輪生於莖頂。花單一，自莖頂抽出，花被兩輪，外輪披針狀，黃綠色，內輪絲狀，造型甚為奇特，花期為4~6月。性喜陰涼，台灣冷杉林下、灌叢下或箭竹林內可見之。

小喜普鞋蘭（小老虎七）

Cypripedium debile Reichb.f.

蘭科 Orchidaceae

多年生小型地生蘭，莖單生，直立。葉2枚，對生，著生於莖頂，圓心形。花由莖頂抽出，單生、下垂，整體黃綠色，唇瓣袋形，內側有紫黑色條紋。性喜陰涼略溼潤之處，中、高海拔之針葉樹林下偶可見之。

攝影／賴國祥

錫杖花

Monotropa taiwaniana Ying

水晶蘭

Cheilotheca humilis (D. Don) H. Keng

鹿蹄草科 Pyrolaceae

皆為多年生依菌生植物，本身無葉綠素，無法行光合作用,需靠吸收菌類分解有機物之養分而存活。錫杖花植物體黃色，葉退化為鱗片狀。花黃色多數頂生，鐘形下垂。性喜陰涼略濕之環境，5月下旬出土至8月於台灣冷杉林下偶可見之。水晶蘭植物體白色，花單一頂生，白色或稍帶粉紅，玲瓏可愛，其柱頭紫黑色，花絲有毛。另有阿里山水晶蘭（C. *macrocarpa* (H.Andres) Y.L.Chou）柱頭白色，花絲光滑。其花期皆較錫杖花早約 1 個月。

■錫杖花。

■水晶蘭。

■阿里山水晶蘭。

玉山鬼督郵（高山兔兒風）

Ainsliaea reflexa var. *nimborm* Hand.

菊科 Compositae

多年生小草本，根生葉叢生。5~7月開花，花冠白色，數朵著生於伸長之花莖上。性喜陰涼之處，常見之於台灣冷杉林下或林緣空曠處。

台灣劉寄奴

Nemosenecio formosanus (Kitam.) B.Nord

菊科 Compositae

多年生草本，莖單獨直立。葉一回羽狀複葉，表面綠色，背面蒼白有絨毛。頭狀花序黃色，多數著生莖頂，「花」雖小，然朵朵清楚可辨，花期8~9月。性喜濕潤而略有陽光之處，台灣冷杉林下、山溝、溪谷兩側偶可見之。

攝影／賴國祥

台灣貓兒眼睛草

Chrysosplenium lanuginosum Hook .f. & Thoms. var. *fosmosanum* (Hayata) Hara

虎耳草科 Saxifragaceae

多年生匍匐小草本，植株小，然常群聚生長，全株有毛，葉互生，近圓形。花頂生，甚小，「瓣」綠4裂，於3~6月開花。另有大武貓兒眼睛草（C. *hebetatum* Ohwi）亦全株有毛，然葉對生，「瓣」白4裂，於4~6月開花。兩者皆性喜陰溼，於台灣冷杉林下或山溝、溪谷旁可見之。

■台灣貓兒眼睛。

■大武貓兒眼睛草。

台灣纖花草

Theligonum formosanum (Ohwi) Ohwi & Liu

纖花草科 Theligonaceae

多年生小草本，莖基部多分枝，匍匐而斜上，全株多少被有柔毛。葉具柄，三角形至卵圓形，由下而上，從對生至互生。花單性，著生於節上，雄花較顯著，花藥線形，花絲纖細、下垂，花期5~7月。性喜陰濕，於台灣冷杉林下或山溝、溪旁可見之。目前之採集記錄僅見於北大武山及合歡山。

延齡草

Trillium tschonoskii Maxim.

百合科 Liliaceae

多年生草本，莖單出直立，葉三枚輪生，著生於莖頂，闊心形。花單朵頂生白色，著生於三片葉子中間，甚為秀麗，花期為3~5月。性喜陰濕，常見於台灣冷杉林下及山溝兩側。

玉山杜鵑

Rhododendron pseudochrysanthum Hayata (Yushan Rhododendron)

杜鵑科 Ericaceae

常綠灌木或小喬木，葉長橢圓形，全緣。盛花期4~5月，花色多變，常由花苞之鮮粉紅色至盛開之白色，在陽光照射下，透著絲質光彩，氣質非凡，更增高山丰彩。廣泛分布於台灣3,000公尺以上之區域，常與香青群聚生長於各峰頂或山脊稜線，構成台灣高山灌叢景觀。

攝影／賴國祥

大花雙葉蘭

Listera macrantha Fukuy.

梅峰雙葉蘭

Listera meifongensis H.J.Su & C.Y. Hu

南湖雙葉蘭

Listera nankomontana Fukuy.

玉山雙葉蘭

Listera morrisonicola Hayata

蘭科 Orchidaceae

四種皆小型地生蘭類，莖纖細，直立或斜上，葉闊半圓形著生於莖之中部，僅兩葉對生。花2~6枚著生莖頂，除玉山雙葉蘭為黃綠色外，餘為淡綠色。四種之區別在於花的構造，尤其是唇瓣的形狀，大花雙葉蘭唇瓣圓鈍、寬闊，邊緣透明；梅峰雙葉蘭唇瓣闊長形，邊緣波浪狀，而且花柄與花柱構成一圓弧狀；南湖雙葉蘭唇瓣先端狹長，分叉明顯；至於玉山雙葉蘭之唇瓣除顏色為黃綠色外，質地較堅厚，且為長方形。四種雙葉蘭之花期雖稍有差異，然大致為6月底至7月。除玉山雙葉蘭生長於陰涼之灌叢下外，餘三種皆性喜陰濕，台灣冷杉林下、山溝、溪谷兩側或箭竹林下可見之。

■南湖雙葉蘭。

■大花雙葉蘭。

■梅峰雙葉蘭。

■玉山雙葉蘭。

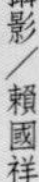

攝影／賴國祥

台灣金蓮花（金梅草）

Trollius taihasenzanensis Masamune

毛茛科 Ranunculaceae

多年生直立草本，莖光滑，約30cm高。莖生葉具長柄，掌狀5~7裂。花金黃耀眼，常單獨頂生，於6月可見。性喜潮濕而陽光適中處，目前僅發現於大霸尖山、奇萊山與合歡山，應多加珍惜、保護。

雙黃花堇菜

Viola biflora L.

堇菜科 Violaceae

多年生小草本，葉厚，腎形，鈍鋸齒緣，多叢生基部。花單朵黃色，由葉基抽出，瓣5枚，部分具紫褐色條紋，花期為6~7月，然開花時日甚短。性喜濕潤而陽光適中處，群聚生長於海拔3,100~3,500公尺間之灌叢下或台灣冷杉林下。

玉山箭竹

Yushania niitakayamensis (Hay.) Keng. f.（Yushan Cane）

禾本科 Gramineae

多年生竹類，一般林外常見者約30~60公分，然林內可達5公尺高，以地下莖拓殖生長，單桿直立，罕開花，為台灣高山草生地之主要組成分子。分布廣泛，台灣亞高山地區從林下一直到稍有土層之處皆可見之。

反捲葉石松

Lycopodium quasipolytrichoides Hayata

石松科 Lycopodiaceae

小型地生蕨類，直立，小葉約1公厘寬，明顯朝下反折，孢子囊黃色長於葉腋。群聚生長於陰溼之台灣冷杉林下或溪旁。常綠性，全年可見。

攝影／賴國祥

生命的韌性

高山寒原

Alpine Tundra

◎撰文／王震哲
◎ Text ／ Jenn-Che Wang

高山寒原係指林木界線以上之植被類型，生育於高山寒原的植物即稱爲「高山植物」。高山寒原植被的特徵爲植物呈塊狀叢生，覆蓋常不連續，組成之種類以草本爲主。由於受到高山地區嚴苛環境因子的限制，生育其間的植物往往具有特殊之形態適應。台灣高山寒原植物所屬的科屬絕大部份均與北溫帶地區相同，並無台灣特有屬，但特產種的比例則甚高，此與過去數百萬年來的冰河週期有密切的關係。不論就生態、演化、或生物多樣性的角度來看，台灣的高山寒原具有無可取代之珍貴價值，誠爲台灣植物基因之寶庫。

The term "alpine tundra" refers to the vegetation above the timber line. Any plants that grow in the alpine tundra areas are known as "alpine plants." The vegetation of the alpine tundra grows in patches, and frequently does not have continuous ground coverage. The plant species here are primarily herbs. Since these plants are limited by the severe environmental factors, they often adapt by developing special shapes. Taiwan's alpine tundra plants are mostly the same genera seen in the northern temperate regions, without any endemic genus to the island. However, there is a high proportion of endemic species. This is intimately related to the glaciations that occurred several million years ago. Whether in terms of ecology, evolution, or biodiversity, Taiwan's alpine tundra has an irreplaceable value, and has become a genetic treasury for Taiwan plants.

■遠眺雪山北側圈谷，綠色部份為玉山圓柏和玉山杜鵑組成之矮盤灌叢，灰色部份為岩屑地，零星散生草本植物，為台灣寒帶植物基因之寶庫。

攝影／王震哲

■玉山之森林界線。

攝影／呂玉娟

當我們在北半球由赤道向北極旅行時，常可發現森林逐漸由闊葉林轉變為針葉林，再往北大約在極地附近即轉變為草本植物為主之植被，此一明顯轉變之界限為森林分布之最北界限，稱之為林木界限（timber line），在林木界線北方以草本植物為主之植物社會即稱為寒原（tundra）。寒原之下層土壤形成永凍層，故又稱凍原，受此因素之限制，故植物的種類以草本及苔蘚類為主。同樣當我們在觀察熱帶山區植物之垂直分布時，也可以發現有些山峰之森林僅分布至某一高度，再往上即轉變為灌木或草本植物為主，類似寒原之植物社會。此一界限在台灣約見於海拔3,500~3,600公尺左右，尤以玉山、雪山及南湖大山最為明顯。此一海拔高度以上即屬高山寒原（alpine tundra）植被類型，亦有學者將其稱為高山岩原及岩屑地。生育於高山寒原的植物即稱為「高山植物（alpine plant）」。

高山寒原植被的特徵為植物呈塊狀叢生，覆蓋常不連續，組成之種類以草本為主，木本植物之種類甚少，即有亦以灌木為限。木本植物多出現於地勢較平坦，土壤堆積較厚之處，往往形成低矮匍伏狀之濃密灌木叢，在生態學上稱為矮盤灌叢（Krummholz），主要的種類為玉山圓柏（*Juniperus squamata* var. *morrisonicola*（Hayata）Li & Keng）及玉山杜鵑（*Rhododendron pseudochrysanthum* Hayata），常成大片叢生。草本植物多散生於灌木叢之間，灌叢之下層因避風且土壤與水分等環境因子較佳，故一些能耐陰之種類亦見於此。

嚴苛的環境
與特殊的適應方式

高山寒原的環境因子相當嚴苛，嚴重限制了植物的生長，唯有具特殊適應能力的植物才能存活，因此高山寒原的植物相較之中低海拔山區顯得相當單調。生育其間的植物往往具有特殊之形態，以適應高山地區強日照、低溫、風大、土壤瘠薄等嚴酷之環境。

一、低溫

山區氣溫隨海拔高度上升而遞減，根據玉山北峰測候所（海拔3,850公尺）之資料顯示各月平均溫度均在10℃以下，年平均溫僅3.7℃，全年有6個月之溫度低於5℃；年雨量約3,000公厘，平均雲天日數25日，冬季有雪，每年12月至翌年2月為積雪期，氣候特徵為寒冷潤濕。由於冬季低溫及積雪的限制，使得植物的生長期間大為縮短，常僅約半年之久即完成其生活史。故生育其間的植物常為一年生的物種，以種實度過寒冬，如毛茛屬（*Ranunculus*）、龍膽屬（*Gentiana*）、佛甲草屬（*Sedum*）等；一些多年生之草本植物則其地上部份在冬季枯萎，以地下之根、莖度過嚴寒的冬季。每年春季積雪初融，高山植物即迅速萌發生長；夏季為高山的花季，由於花期集中，再加上高山植物多半花大且顏色鮮豔，百花齊放，美不勝收；花後隨即結果，迨10月霜降，種實已飛散休眠。

■高山植物花大且顏色鮮豔。奇萊紅蘭（*Orchis kiraishiensis*）

攝影／王震哲

二、日照強烈

高山上空氣較為稀薄，日照較平地強烈。且因陽光穿過大氣層時不同波長色光之穿透力並不相同，波長較短的紫外光其穿透力較弱，故高山上紫外光之強度比平地強。高山植物為避免因強烈紫外光所造成的危害，常產生大量的類黃素來吸收過量的紫外光。而類黃素為植物花色的主要成分之一，因此高山植物之花色常較為鮮豔即肇因於此。

三、水分缺乏

含有水汽之氣流受地形抬升之後，因溫度降低而凝結降雨，因此山區的雨量常隨海拔上升而遞增。但當海拔超過一定高度後，因空氣中之水汽已大部份凝結析出，此時降雨就會隨著海拔上升而遞減，故高山寒原的雨量常較其下方之針葉林帶為低。且因高山之地形陡峭，土壤化育不佳，而無法保留水分。故高山植物之葉片常為肉質或革質、或葉面生長毛茸等，以減少水分蒸散。

■南湖大山圈谷生育著多種台灣高山特有植物。 攝影／王震哲

四、風力強勁

■為適應高山的乾旱環境，高山植物葉片變肥厚。玉山佛甲草（*Sedum morrisonensis* *Hayata*）。 攝影／王震哲

高山上受地形影響，風力常極為強勁。植物受強勁風力吹襲以致無法直立生長，最明顯的例子如玉山圓柏，此種植物在避風處，如雪山的翠池以及南湖圈谷，可長成直立之喬木，但在山頂或稜線附近則成匍伏生長；草本植物則多匍伏生長，或植株低矮，或生育於避風之處。此外，強勁風力帶走土壤使得高山土壤瘠薄，風也會加速植物的蒸散作用，故生育於強風吹襲環境下之植物常具有類似乾生植物之形態。

冰河孑遺種

若仔細分析生育在台灣高山寒原之植物種類，可以發現這些種類所屬的科屬絕大部份均與北溫帶地區相同，其中並無台灣特有屬。例如常見之高山植物細葉薄雪草（*Leontopodium microphyllum* Hayata）屬菊科之薄雪草屬（*Leontopodium*），此屬約有30種，廣泛分布於歐亞大陸之溫帶至寒帶地區，其中最廣為人知的種類即為分布在阿爾卑斯山的Edelweiss。其他在高山寒原較常見的屬如沙參屬（*Adenophora*）、龍膽屬（*Gentiana*）、櫻草屬（*Primura*）、佛甲草屬（*Sedum*）、籟蕭屬（*Anaphalis*）、柳葉菜屬（*Epilobium*）、羊茅屬（*Festuca*）、髮草屬（*Deschampsia*）...等，亦均為廣泛分布於北溫帶地區的植物。

何以位居亞熱帶地區的台灣，其高山上的植物竟與相距數千公里的溫帶地區如此相似？此與過去數百萬年來的冰河週期有密切的關係。由於台灣島的形成大約是在二百萬年至三百萬年前，其後又因冰河作用而數度與鄰近的陸地相連，故我們可以推測，在台灣島形成之初，植物利用不同的方式由鄰近地區播遷到台灣。冰期的時候，因溫度下降，北方的植物向南遷徙到達台灣，到了間冰期，由於溫度上升，溫帶植物又逐漸向北退卻，但由於台灣地勢高聳，故有些種類可退至高山地區殘存。在這些退至高山地區的種類中，有些種類或因演化速率較為緩慢，以致仍維持與北溫帶地區或熱帶高山地區相同之種。如雙黃花菫菜（*Viola biflora* L.）即為一例，此種植物在台灣僅零散分布於玉山、雪山、南湖大山、合歡山等高山地區，數量相當稀少。但在國外其他地方則廣泛分布於整個北溫帶地區，甚至在熱帶地區之婆羅洲高山上亦可見之，族群呈現退縮隔離之分布類型。其餘如小杉葉石松（*Lycopodium appressum* Desv.）、扇羽陰地蕨（*Botrychium lunaria* (L.) Sw.）、高山梯牧草（*Phleum alpinum* L.）、綠花凹舌蘭（*Coeloglossum viride*（L.）Hartm.）...等亦呈類似之分布情形。這類植物由於曾經歷過冰河作用而仍留存下來，故常被稱為冰河孑遺植物。

■雪山主峰附近的高山寒原，由於受強風吹襲影響，植物長成低矮匍伏之矮盤灌叢，以玉山圓柏和玉山杜鵑為主。 攝影／王震哲

上■玉山石松（*Lycopodium veitchii*）間斷分布於中國西南及台灣，為冰河孑遺植物。 攝影／張和明

下■南湖大山柳葉菜（*Epilobium nankotaizanense*）為依據文化資產保存法指定的珍貴稀有植物。 攝影／陳志雄

植物基因的寶庫

但若以種來看，則有頗多為台灣之特產種（endemic）。在1937年，日本的植物分類及生態學大師正宗嚴敬曾針對台灣3,000公尺以上高山之維管束植物進行調查，共列舉了373種，其中有243種（約

65%）為台灣之特有種，特有種之比例較之全島植物特有比例（約25~40%）顯然高出甚多。這些種類僅生育於台灣，但與溫帶其他地區之種類極為類似或具有極相近之親緣關係，其同屬之近緣種亦都見於溫帶地區。這些特有種之物種形成機制為生物演化上相當有趣之問題，一般而言，屬的演化歷史需要較長的時間，如前所述，台灣島與鄰近地區之隔離時間還不足以形成特有屬；但種的演化則較為快速，推測這些台灣特有種種類大部份為間冰期退至高山之後仍繼續其演化過程，且因與其他地區之族群有地理上之隔離，無法交換遺傳物質，經過漫長歲月，遂在台灣演化出獨立之新種。例如最近發現之奇萊肋柱花（*Lomatogonium chilaiensis* C. H. Chen & J. C. Wang），為侷限分布於奇萊山之台灣特有種，與其親緣關係最密切之種類為廣泛分布於歐亞溫帶地區之肋柱花（*L. carinthiacum*（Wulfen）Reichenbach）。這些特有植物率多以台灣之高山地名命名，如阿里山龍膽（*Gentiana arisanensis* Hayata）、奇萊青木香（*Saussurea kiraisanensis* Masamune）、奇萊紅蘭（*Orchis kiraishiensis* Hayata）、玉山小蘗（*Berberis morrisonensis* Hayata）、玉山飛蓬（*Erigeron morrisonensis* Hayata）、玉山茴芹（*Pimpinella niitakayamensis* Hayata）、玉山佛甲草（*Sedum morrisonensis* Hayata）、南湖附地草（*Trigonotis nankotaizanensis*（Sasaki）Masamune & Ohwi *ex* Masamune）、南湖大山紫雲英（*Astragalus nankotaizanensis* Sasaki）、南湖毛茛（*Ranunculus nankotaizanus* Ohwi）、雪山馬蘭（*Aster takasago-montanus* Sasaki）、雪山堇菜（*Viola adenothrix* var. *tsugitakaensis*（Masamune）J. C. Wang & T. C. Huang）、雪山翻白草（*Potentilla tukitakensis* Masamune）、...等等，都僅見於台灣的高山寒原，不論就生態、演化、或生物多樣性的角度來看，均具有無可取代之珍貴價值，誠為台灣植物基因之寶庫。

上■高山寒原環境嚴苛，冬有積雪，植物常為一年生。高山毛茛（*Ranunculus junipericolus*）

下■奇萊肋柱花（*Lomatogonium chilaiensis*）為台灣特有種，目前僅知分布於奇萊山區之高山岩屑地。

攝影／王震哲

九命怪貓

箭竹

Yushania niitakayamensis

◎撰文／韓中梅

◎ Text ／ Chung-May Han

台灣地區許多山頂都是大片的草原，在高海拔地區以玉山箭竹形成的箭竹原最爲普遍，這種大面積的箭竹原是森林大火之後的景觀。通常林下生長的玉山箭竹高可達兩公尺以上，位於空曠處的族群有矮化至40公分甚至更矮的現象。箭竹是以地下莖進行無性繁殖爲主要拓殖方式，在森林火災之後，潛伏地下的竹鞭提供了新芽快速抽發的保障，因此在多次林火肆虐後，玉山箭竹仍可迅速的恢復翠綠舊觀。而在低海拔的陽明山地區則是包籜矢竹的地盤，和玉山箭竹一樣，也是森林消失後的優勢種類。包籜矢竹在2000年出現大規模開花，也使得台灣對包籜矢竹的研究展開一個新的里程。

Many mountain peaks in Taiwan are covered with large expanses of meadows, and in high-altitude areas like Jade Mountain, meadows composed of *Yushania niitakayamensis* are common. These large expanses of *Yushania* are the scene after forests have been destroyed by fire. Very often, the *Yushania* that grows under a forest can reach two meters or more in height, and clumps in open areas can be as short as 40 centimeters or even less. *Yushania* reproduces primarily asexually, by means of underground rhizomes. After a forest fire, the shoots buried deep underground are like a guarantee, quickly sending up new runners, for which reason after a forest has been hit by fire, *Yushania* can rapidly restore a green visage to the landscape. And on sunny mountain slopes at low-altitudes, *Pseudosasa usawai* basins are seen, and like the *Y. niitakayamensis* of Jade Mountain, this is also the predominant species seen after a forest has been destroyed. *P. usawai* had a large-scale bloom in 2000, which marked a new milestone in research into this species in Taiwan.

■玉山箭竹形成的高山草原。看似柔軟的草坡其實是玉山箭竹地下莖強韌生命力的表現。 攝影／韓中梅

台灣地多高山，在山頂常見大片蔥綠草原，這種高山草原常是由兩大要角所組成 - 玉山箭竹（*Yushania niitakayamensis* (Hayata) Keng f.）與芒草。在海拔 1,500 公尺以上的山區即可見玉山箭竹在針葉林下形成大面積族群，上升至海拔 3,000~3,500公尺的地區，尤其是接近山頂部分，更是形成大片的草原。在橫跨海拔 2,000 公尺的廣大生態幅度中，外型變異極大，在林下生長的玉山箭竹往往可以高達兩公尺以上，穿梭鬱閉的箭竹叢中常有只聞聲不見人的情形，而隨海拔增高，位於山頂等風力強勁處或是林火、山崩頻繁致使土壤肥力貧瘠地區的族群有矮化至40公分甚至更矮的現象。

玉山箭竹是植物界的九命怪貓。因其以地下莖進行無性繁殖為主要拓殖方式，在森林火災發生，地表植株燒毀的情況下，潛伏地下的竹鞭提供了新芽快速抽發的保障，往往在多次林火肆虐後，玉山箭竹仍可迅速的恢復翠綠舊觀。是以在森林火災造成之白木林下，常是玉山箭竹群落最後發展成當地天然植被之唯一主要份子，形成玉山箭竹草原。以合歡山主峰為例，經過多次火焚後，白木林僅剩零落枯立白木，現更不復存有灰白枯木，轉變成為玉山箭竹草原。

玉山箭竹另一特殊之處，除了屬名及種小名均以玉山為名外（*Yushania*—玉山，*niitakayamensis*-

■針葉林下的玉山箭竹，體型顯然較山頂的同伴高大許多。

攝影／韓中梅

■玉山箭竹之地下莖，橫走甬匐擴張勢力範圍，纏繞糾結固著已有地盤，無怪縱橫台灣三千公尺以上的高地。

攝影／韓中梅

■回祿造訪後，曾經蓊鬱的常綠森林只留下白木林作為存在的證明，玉山箭竹取而代之成為主要的組成物種。　攝影／韓中梅

新高山，即玉山），也是耿伯介先生在1957年所創之新屬—玉山箭竹屬的模式種，更是竹亞科植物中唯一以台灣物種為名的屬，代表了玉山箭竹在竹類植物分類上的重要地位。

由於地下莖生長旺盛，每年4~6月間玉山箭竹發筍時，常可見黃鼠狼等動物之排遺，也可在玉山箭竹筍上發現被咬的痕跡。除鼠輩外，人類摘採活動亦已行之有年，或為盤中佳餚，或罐裝出售，即所稱之雲筍。因其主枝挺直堅韌的特性，原住民族常用以狩獵，一端削尖做為弓箭素材，兩端削尖則可用於布設箭竹陣，此外亦廣泛用於住家建材等方面。

在低海拔山區，包籜矢竹（*Pseudosasa usawai* (Hayata) Makino & Nemoto）是台灣另一草原物種，亦是台灣特有種，主要分布於台灣北部，如陽明山及瑞濱臨海山丘稜線等受強烈東北季風作用的地區。與玉山箭竹一樣，包籜矢竹面對不同的生育環境時，也表現出差異極大的外型。山麓地帶多分布於闊葉林下，個體較高，葉片、分枝數都較多，位於山頂則形成純林狀態，植株較矮。以生長在瑞濱的族群為例，位於溪谷背風處的個體高度多半超過兩公尺，而在稜線直接受海風吹拂的個體則矮化至一公尺左右。

自包籜矢竹發表（Hayata, 1916）以來至2000年始有大規模開花之記錄，足見其生活週期甚長。包籜矢竹在族群進行大規模開花結實後即行枯死，也形成陽明山區各山頭大片枯黃之百年難得景象。矢竹林下因大量落種而鋪滿厚實竹米，實為枯黃族群更新復舊之希望所在。當然箭竹種子—竹米是生殖的利器，它萌芽而成綠油油一大片箭竹苗，有的地方一平方公尺可以算到二千餘株，生長雖然快速，半年可以長一尺，可是生存的競爭也非常激烈，死亡率有百分之九十以上，最後能存活的也不過十來株，這十來株要生長到第五年以後才能成叢，才有能力發新筍。此一結果與國外學者對中國大陸臥龍熊貓保護區中箭竹（*Sinarundinaria fangiana* (A.Camus) Keng f. et Wen.）大開花後所做之追蹤結果相同。

左■結實纍纍的種子穗，飽滿的竹米是包籜矢竹延續生機的命脈所在，卻也是鼠輩飽餐一頓的大好時機。

右■幸運逃過鼠輩魔掌及病蟲害的竹米無不努力竄高，彼此競爭的結果是竹苗的數量漸漸減低，趨近母族群的原有密度。

攝影／韓中梅

箭竹為什麼要這麼久才開花，這個問題長久以來吸引著大家的注意。密西根大學的簡生（Daniel H. Janzen）博士對這個現象提出了假說。他認為這和吃竹子種子的動物數量變化有關係。竹子大開花就結了大量的竹種子，又稱竹米。竹米沒有毒，比稻、麥更有營養，大量落在地上，竹叢之下常呈厚厚的一層。美味的竹米會吸引鳥類、老鼠還有人類的注意，許多動物的族群大小都因為這百年大事而劇烈的波動。

上■包籜矢竹在陽明山區也形成大片的純林，頗有草原的氣勢。

下■1999-2000年的大規模開花使得終年常綠的陽明山像覆雪似的枯黃了山頭。

攝影／韓中梅

除了生態學，包籜矢竹的開花也引起學者的高度興趣。由於缺乏大開花的紀錄，過去認為包籜矢竹的基因變異，僅由不斷的分櫱生殖累積而來，累積的突變無法藉有性生殖在族群內及族群間流傳，只能隨著營養系擴大，族群內及族群間應有高度的遺傳分化。然而根據研究，包拓矢竹族群內的遺傳變異豐富，遺傳物質在族群及亞族群間的交流並未受到明顯阻礙，當時即推測包籜矢竹除了利用每年發筍進行無性生殖以外，開花結實必然也扮演了重要的角色（陳等，1998）。這一個推測在當時沒有任何的證據可以支持，從包籜矢竹正式發表以來文獻中沒有任何大規模開花紀錄。及至1999~2000年陽明山區族群進行大規模開花才首度獲得證實。該研究也發現陽明山的中湖及小油坑亞族群在遺傳組成上與金瓜石族群較為接近，鄰近中湖及小油坑的大屯山亞族群反而在遺傳上並非最接近的。另一研究也顯示相近的結果，小油坑、中湖、金瓜石及瑞濱等地的族群在親緣上較為接近，甚至可以推論與大屯山區亞族群間以久未有基因流傳的發生。從其他的種種蛛絲馬跡可以看出小油坑、中湖與金瓜石、瑞濱的族群雖然目前的範圍沒有相接，但是在過去應該是相互連接的。更進一步的分析顯示這些族群是由小族群歷經快速的範圍擴張而來，甚至可以推斷金瓜石族群即為前述之小族群，如果配合台灣過去的地質歷史來看，金瓜石族群可能為冰河時期包籜矢竹族群退縮的避難地所在，目前瑞濱、中湖及小油坑等地個體是冰河後退、氣候暖化後自金瓜石重新擴展而來。

沒有美麗的花朵，玉山箭竹與包籜矢竹在不同的海拔用相似的方式悄悄地站穩在台灣的山頭。

高山仙子

高山野花

Alpine Wildflowers

◎撰文／呂勝由、鄭育斌

◎Text／Sheng-You Lu
Yu-Pin Cheng

位於亞熱帶氣候區的台灣，具有複雜多變的地形，提供多樣的棲地，形成高度歧異的物種。台灣的高山隨著四季變化而有不同的面貌，豐富的植物組成更是台灣珍貴的自然資源。美麗的高山野花，是高山生態系中最動人的景緻。高山的氣候嚴峻、土壤淺薄，對植物而言是極具挑戰的生育環境。生長於高山地區的美麗野花僅能利用一年中短暫的生長季節，來進行開花結果等生長。爲了適應此特殊的氣候環境，在外部型態和內在的生理上產生了特化，多毛、葉厚、反捲，地下根莖部位膨大等。爲了對抗強烈的紫外線，花青素特別發達，使高山植物呈現豐富而鮮豔的顏色變化。許多的台灣高山植物都屬於冰河孑遺物種，在最後一次冰河期結束後，這些物種往高海拔遷徙，殘存於高海拔山區。每一座獨立的山峰都像是島嶼般的，將族群隔離，造成族群間的基因交流困難。隔離與特化的結果，讓台灣的高山植物具有豐富的物種組成和高度的特有性。

Taiwan is positioned in a subtropical location, and has a complex and varied topography. This provides a variety of habitats and means that Taiwan's plants are highly differentiated. Taiwan's alpine regions have a different face as the seasons come and go. The rich plant life here forms a rich natural resource for Taiwan to treasure. The beautiful alpine wildflowers provide the most moving spectacle of the alpine environment. The severe climate and thin soil of these mountain regions means that plants find it a very challenging environment in which to live. The beautiful wildflowers that grow in high mountain regions make full use of the short annual growing season to flower and set fruit. In order to adapt to the special climactic conditions here, these flowers are specialized in external and internal structure. They have cilia and thick leaves, and tend to curl. Their root systems below ground cover a great area, and in order to counteract the effects of ultraviolet light, the chlorophyll in these plants is particularly well developed. This means that the alpine species display rich and vibrant color changes. Many of Taiwan's alpine plants are relic species from the ice ages. After the disappearance of the last glaciers, these plants moved to the higher altitude areas where they are still found today. Each individual mountain peak seems like an island, and species are isolated. It is difficult for genes to flow between communities. The result of this isolation and the specialization of the plants is to bring a rich diversity and a high degree of uniqueness to Taiwan's alpine plants.

■南湖大山高山景緻。

攝影／呂勝由

位於亞熱帶氣候區的台灣，島內由幾條主要的南北向山脈，構成了台灣島的主體。島內的地形複雜多變，提供了多樣的生育地，從而孕育了台灣多樣的物種和生態系。由於位於太平洋海板塊和歐亞大陸板塊交會區，持續不斷的板塊擠壓和旺盛的造山運動，使得台灣島內高山林立，3,000公尺以上的高山就超過200餘座，最高峰的海拔近4,000公尺。台灣的緯度雖然不高，但因地形起伏，島內的植群垂直分化明顯，在高海拔山區主要是以針葉林為主，針葉林以上的森林界限，則產生類似於極區的高山寒原景觀。狹義的高山植物，主要是指分布於森林界限以上的植物，廣義的高山植物則是泛指生長於高海拔區域的植物。當然在高海拔區域主要的植群是針葉林，但是在針葉林下、高山草原中，或是森林界線以上的一些高山野花，會在不同的季節，或獨自綻放、或形成大片群落，往往會形成令人讚嘆的美麗景觀。

高山的四季並不像平地般的明顯，春季總是來的特別遲，在暮春3月時，平地早已是春暖花開，這時候高山的野花才剛自冰冷的大地中甦醒，開始開枝展葉，啟動一年中的生長過程。冬季也是來得特別早，大約在10月中下旬便進入冬季，冬季來臨之後植物的生長停滯、景緻單調，整個高山上的野花銷聲匿跡。一般而言，夏季是高山植物最主要的生長季節也是最美的季節，也是高山地區最美的季節。繽紛絢爛的花季大約四月中以後就開始，由2,000多公尺的山區，往上蔓延至近4,000公尺的山區。四月底開始的高山杜鵑花季，梅雨季期間，森氏杜鵑(*Rhododendron pseudochrysanthum* Hayata ssp.*morii* (Hayata) Yamazaki）、玉山杜鵑（*R. pseudochrysanthum* Hayata)在雨霧中開遍了山區。粉紅的紅毛杜鵑（*R. rubropilosum* Hayata）在6月份像野火般的開滿整片的高山箭竹草原，潔白芳香的台灣百合（*Lilium formosanum* Wallace）也漫山遍野的盛開著。高山的花季在7、8月份達到高峰，野地裡呈現高度色彩歧異度的龍膽類（*Genetina spp.*）植物，藍紫色的玉山水苦賈（*Veronica morrisonicola* Hayata），鮮黃的毛茛類（*Ranunculus spp.*）、玉山金絲桃（*Hypericum nagasawai* Hayata）和玉山佛甲草（*Sedum morrisonense* Hayata），雪白的玉山薔薇（*Rosa sericea* Lindl. var. *morrisonensis*（Hayata）Masamune）、高山薔薇（*R. transmorrisonensis* Hayata）、梅花草（*Parnassia palustris* L.）、白花香青（*Anaphalis morrisonicola* Hayata）等植物熱熱鬧鬧的開遍高山的每一片草原和岩屑地。進入秋季之後，高山呈現不同的景緻，在入冬之前，氣溫漸降，植物體內色素體產生變化，所以這個季節的高

■藍色系花的玉山水苦賈。 攝影／呂勝由

■黃色花的玉山金絲桃。 攝影／呂勝由

■花色雪白的高山薔薇。 攝影／呂勝由

■花色雪白的玉山薔薇。 攝影／呂勝由

上■生長在岩屑地的植物，具有粗大主根系的玉山櫻草。

中■具肉質的葉片的玉山佛甲草。

下■植物體毛茸茸的尼泊爾籟簫。 攝影／呂勝由

■深根性的玉山當歸。 攝影／呂勝由

山景緻以黃或紅的葉色變化為主。冬季的高山，低溫和冰封，使植物的生長停滯，萬物都在休眠中期待下一個春天的來臨。

高山地區的氣候環境非常的多變，氣溫偏低，一年之中適合生長的季節非常短暫，加上土壤發育不良，強風暴雨等特殊的環境因子，對植物來說都不是好的生長環境。為了適應高山寒冷的氣候和貧瘠的土壤，高山的野花各自演化出特殊的形態特徵來適應惡劣的環境。例如生長在岩屑地的植物，就具粗大的主根系，用來固著植株和吸收土壤中的水份和養份，以高山山蘿蔔（*Scabiosa lacerifolia* Hayata）、玉山櫻草（*Primula miyabeana* Ito&Kawakami）和玉山當歸（*Angelica morrisonicola* Hayata）等最具代表性。葉為植物最主要的營養器官和光合作用的部位，為了減少體內水份的散失，和進行有效的光合作用，許多高山植物的葉肉特別的肥厚，內部的柵狀組織特別發達，景天科植物，如玉山佛甲草、火焰草（*Sedum stellariaefolium* Franch.）等的植株都是屬於肉質的葉片。除此之外，葉表密被蠟質、莖葉上密生茸毛或是葉緣明顯的反捲，這些特化的構造都是為了適應高山地區惡劣的氣候環境，達到延續生命的目的。白花香青、尼泊爾籟簫（*Anaphalis nepalensis* (Spreng.) Hand.-Mazz）、玉山金梅（*Potentilla leuconota* D. Don var. *morrisonicola* Hayata）和雪山萎陵菜（*P. tugitakensis* Masamune）等，在植物體或是葉外部具有茸毛。高山的杜鵑類植物，如玉山杜鵑和森氏杜鵑，通常都具有蠟質的葉表和毛茸茸的葉被，葉緣部位反捲。由於生長季節短暫，高山野花為了延續生命，會採用不同的策略來度過寒冷的冬季。多年生的植物，在冬季地上部枯萎，僅剩粗大的地下莖或根部度冬；一年生的植物，則是利用一年中短暫的生長季節，快速的生長，開花結果，完成生活史，留下種子待來年重新萌發，來延續族群的生命。事實上，由於氣候條件太過於惡劣，一年生植物的數量非常少，大多數植物都採多年生的生長模式，來確保族群的延續。

陽光是植物生長的必需和能量的來源，但是太強的光照對於生長在高山地區的植物而言反而是一種困擾，除了加速水分的蒸散之外，陽光中的紫外線、紅外線，對植物的生長會造成抑制作用。為了保護自已，高山植物的花或葉通常會含有較多量的花青素，用來吸收過量的紫外線。這些花青素除了具有保護植物的功能之外，也讓至高山花卉更加的美麗動人，大多顯現出紫、紅、黃等鮮豔的顏色。例如，藍紫色的玉山沙參（*Adenophora morrisonensis* Hayata）、高山沙參（*A. morrisonensis* Hayata spp. *uehatae* (Yamamoto) Lammers）、台灣烏頭

上■藍色系花的高山沙參。
中■藍色系花的玉山山蘿蔔。
下■藍色系的高山烏頭。　攝影／呂勝由

（*Aconitum fukutomei* Hayata var. *fukutomei*）、阿里山龍膽（*G. arisanensis* Hayata）、玉山山蘿蔔、高山倒提壺（*Cynoglossum alpestre* Ohwi）等；紅色系列的玉山櫻草、玉山石竹（*Dianthus pygmaeus* Hayata）、玉山蒿草（*Pedicularis verticillata* L.）、疏花光風輪（*clinopodium laxiflorum* (Hayata) Mori var. *laxiflorum*）、早田氏香葉草（*Geranium hayatanum* Ohwi）等；黃色系花的玉山金絲桃、玉山龍膽（*G. scabrida* Hayata）、黑斑龍膽（*G. scabrida* Hayata var. *punctulata* S.S.Ying）、高山毛莨（*R. junipericolus* Ohwi）、疏花毛莨（*R. formosa-montanus* Ohwi）、玉山佛甲草、玉山金梅、雪山翻白草（*P. tugitakensis*

左■開粉紅色花的玉山蒿草。 右上■黑斑龍膽的黃色花瓣上散佈著黑色斑點。 右下■開黃花的疏花毛莨。 攝影／呂勝由

Masamune）、一枝黃花（*Solidago virgaurea* L.var. *leiocarpa* (Benth.) A. Gray）等應有盡有。

位於亞熱帶的台灣，卻在高山的寒原地帶殘存一群與極地寒原有親緣關係的植物，探究這一群植物的來源，就要先了解台灣在過去地質年代的歷史。對台灣的生物分布影響最大的原因，首推第四紀的冰河(glacial)事件，尤其是結束於一萬年前的最後一次冰河，形成目前的隔離狀態。第四紀多次的冰河事件，雖然沒有直接侵襲台灣，但是在大氣候的影響下，氣溫降低、海平面下降，台灣與大陸陸塊間經由陸橋的聯繫，北方極地寒原成分的植物，在冰河時期南遷進入台灣，避開冰河的侵襲。而在冰河退卻、氣候暖化的過程中，這些原本生活於較冷環境的植物，卻因海峽的隔絕重新形成，或因氣溫的快速變遷而來不及向北退卻，為求生存只好選擇向高山地區遷徙。幸運的，台灣地區有高山地帶可供這群植物避難，延續族群的生命；不幸的是，這群植物族群驟縮後，被隔離在各個孤立的山峰上，就如同被隔離在許多的小島般，無法進行族群間的基因交流，誰也無法保證它們能再生存多久。

隔離除了造成遺傳基因的窄化，也造成族群間的遺傳分化，新的分類群就在這種隔離機制下產生。許多的台灣高山植物，在高山生態島嶼化的地形隔離機制之下，為了適應環境逐漸演化成亞種或是特有種。這也是為什麼台灣的高山地區會有這麼多的特有種植物的主要原因。台灣高山地區有名的冰河期孑遺物種，如尼泊爾籟蕭、玉山薄雪草等分布於高山的寒原地帶，南湖柳葉菜僅侷限分佈於南湖大山、奇萊山和關山山頂，碎雪草屬內的幾種植物，則是隔離分布於台灣主要的高山地帶。這些植物雖然都是台灣特有種，但是在溫帶地區或是在喜馬拉雅山區，還是可以找到親緣關係相近的分類群。如玉山薄雪草就和生長於阿爾卑斯山的「小白花」(*Leontopodium alpinum* Cass.）同屬，這些植物的分布為台灣的植物來源或起源留下一些可供追尋的蛛絲馬跡。

■細葉薄雪草。 攝影／呂勝由

台灣的高山美景是令人心動的，但是如果沒有這些野花的點綴，則又略顯單調。當然，台灣的高山地形起伏，想要親近可能得要費一番功夫。玉山、南湖大山等高山是許多人嚮往的高山，但是卻不容易到達。而合歡山區就是一個觀賞高山野花最合適的地點，不僅是交通方便，在這裡大部分的高山野花都能看到。每年6到10月份間，是欣賞高山野花最適合的季節，在這段時間選個晴朗的好天氣，上山去賞花，保證會讓人沉醉在高山美景中，而流連忘返。

覆雪的容顏

菊科植物

The Compositae

◎撰文／彭鏡毅

◎ Text ／ Ching-I Peng

台灣全島山地地形發達，3,000公尺以上的高山超過200座，由於長年風化作用強烈，加上部分地區特殊的地形地質，使得台灣的高山與東部石灰岩區擁有相當多的岩屑地形，生存環境條件嚴苛，土壤稀薄、大雨沖刷、風勢強勁而且冬季積雪，在如此特殊的條件之下，孕育了與衆不同的岩屑地代表性植物，除了衆所周知的禾本科及莎草科植物之外，還有一群另具特色的菊科植物。高海拔岩屑地區的菊科植物與一般園藝觀賞菊花大異其趣，除了花朵豔麗外，葉片或苞片上常密密地覆蓋一層像霜雪般的白色細棉毛，即使在夏季裡，高山上的小野菊也常像是覆了一層薄雪的小白花，在風中輕輕的顫動。

Taiwan boasts a very rugged topography, with over 200 peaks of more than 3,000 meters in elevation. Given the serious weathering, plus the special geographical features of some areas, Taiwan's high mountains and the eastern limestone region are rather given to scree slope. This area is a harsh living environment with scarce soil, heavy rains, strong winds and severe winters. Consequently, to live in these special conditions, a group of special plants came about which are quite different from others. Besides the Gramineae and Cyperaceae with which almost everyone is familiar, there are also a group of very different Compositae. The Compositae of the high-altitude cliff regions are longer in shape than the average Compositae, and apart from the petal-shaped bractioles which replace the function of the tongue-shaped flower, the leaves and bractioles are almost covered with a layer of fine white down that looks like frost. As a result, even if no snow falls in winter, the high-altitude wild chrysanthemums sleep gently onward covered by a layer of white snow, proof against the wind.

■玉山薄雪草具有特殊的輻射狀總苞片，而與葉片都佈滿了細細的白色棉毛，宛如佈滿雪花的一朵小白花。　攝影／彭鏡毅

白花香青（*Anaphalis morrisonicola* Hayata）又名玉山抱莖籟簫，與或稱玉山籟簫的尼泊爾籟簫（*Anaphalis nepalensis* (Spreng.) Hand.-Mazz.）同為菊科籟簫屬植物，是台灣典型的高山植物群之一，兩者都屬於陽性旱生植物，除了廣泛分布全島高山之外，也常成為高山岩屑地的優勢族群。白花香青的分布高度範圍較大，在海拔1,500公尺以上乾燥多石的岩縫或岩壁往往可以找到它成簇的芳蹤，而尼泊爾籟簫的分布則多限於海拔3,000公尺以上，通常成群生長於岩屑地，但族群密度較低，在外型上也相當容易與白花香青區分。根據文獻記載，尼泊爾籟簫曾被譽為「台灣首屈一指的名花」，日治時代數度被印刷在「台灣始政紀念」的明信片上，十分風光。在7月至10月的花期過後，籟簫乾燥的苞片與宿存的瘦果就像是乾燥花般的佇立在寒風，等待來年的春天從岩隙中再抽新芽。

玉山薄雪草（*Leontopodium microphyllum* Hayata）是特產於台灣的高山植物，僅分布在海拔約3,500公尺以上之矮盤灌叢及高山岩屑地，常出現於峰頂、稜線及裸岩的環境。薄雪草屬植物是世界性的冰河孑遺植物，其中以阿爾卑斯山象徵純潔而高貴的小白花（Ederweiss）廣為人知，更以一曲Ederweiss享譽全球。玉山薄雪草頭狀花序的外圍有一輪毛茸茸的葉狀苞片，葉片也佈滿了細細的白色棉毛，宛如佈滿雪花的一朵小白花。

■玉山薄雪草。 攝影／彭鏡毅

在台灣的中海拔山區大約1,200~2,800公尺的地區，偶爾可見到一種像是撐著破雨傘的菊科植物，那就是破傘菊！破傘菊的葉片呈現掌狀深裂，有著長長的葉柄著生於葉片，所以看起來就像是把破傘般在步道旁或森林邊緣隨風搖曳。台灣的破傘菊屬植物有兩種，都是特有種，其一為高山破傘菊（*Syneilesis subglabrata* (Yamamoto & Sasaki) Kitam.），分布於台灣中海拔山區，另一種為台灣破傘菊（*Syneilesis intermedia* (Hataya) Kitam.）。高山破傘菊偶見於森林或林道邊緣，而且族群小、數量不多，屬於IUCN物種保育等級評估中易受害（Vulnerable）之等級。而台灣破傘菊因更因低海拔棲地開發殆盡，日治時代以後，已無採集記錄，可能已經滅絕，也許這就是破傘菊在台灣這塊土地上

■白花香青。 攝影／彭鏡毅

■尼泊爾籟簫。

攝影／彭鏡毅

■高山破傘菊。 攝影／彭鏡毅

的宿命。可以說只有在山林的庇蔭之下，台灣的生物多樣性才能永續發展。

台灣東部石灰岩地區有一種特殊的台灣特有菊科植物，頂著碩大的頭狀花序，它的舌狀花純白、管狀花鮮黃，在岩石坡上組合成了一簇一簇黃白交織的美麗花束，這種只出現在東部石灰岩區的菊科植物叫做森氏菊（*Dendranthema morii* (Hayata) Kitam.）。森氏菊的分布以太魯閣國家公園為主，從海拔400公尺的綠水合流步道至海拔2,408公尺的清水山頂裸岩稜線上皆有零星分布，數量較為稀少，森氏菊具有走莖，在較為開闊的山坡岩塊縫隙中，可見其三五成群的綻放。森氏菊的葉片稍厚，葉下表面佈滿柔軟的銀色棉毛，而頭狀花序更展現中高海拔植物花部比例加大的特徵，使得森氏菊在野外相當容易辨認，是石灰岩地區的代表性植物之一。

在太魯閣地區還有一種極為稀有的菊科黃苑屬（千里光屬）植物，並以太魯閣之名命名為太魯閣千里光（*Senecio tarokoensis* C. I Peng），除了是台灣的特有植物之外，也是僅分布於台灣東部石灰岩區的稀有植物，目前僅在花蓮清水山砂卡噹林道、嵐山及研海林道有採集紀錄，分布高度範圍在海拔1,000~2,000公尺之間；太魯閣千里光在IUCN的等級中屬於瀕臨滅絕（Endangered），族群小、分布狹隘，而值得慶幸的是，太魯閣千里光的分布均位於國家公園生態保護區內，受到良好的保護，唯應控制標本採集的數量，以免對族群數量造成不利的影響。太魯閣千里光葉質稍厚，葉背常為深紫色，並有著黃苑屬植物典型豔黃花色，頭狀花序在台灣產黃苑屬植物中為最大型，舌狀花多達8~12枚，所以開花時期相當耀眼，幸運的話，在林道邊、岩壁坡上可見其蹤跡。

■森氏菊。 攝影／彭鏡毅

■森氏菊。 攝影／彭鏡毅

■太魯閣千里光。 攝影／彭鏡毅

■太魯閣千里光。 攝影／彭鏡毅

淒美堅強的樂章

龍膽

Gentiana L.

◎撰文／陳志雄

◎Text／Chih-Hsiung Chen

台灣的山地陡峭，生態環境變化大，造就了千姿百態的高山植物。這些植物的類群大多起源自溫帶地區，經過向外繁衍、蔓延之後，在這塊美麗的淨土上覓得了適合的環境，並存活下來。部份類群經過長時間的適應與演化後，已經和祖先或親戚有所不同，而成爲台灣地區的特有種；龍膽屬植物就是此類型植物的最佳代表，它們喜好高冷的氣候環境，大多都是一年生草本。這種特性，讓它們可以適應高山地區特別嚴酷的氣候。它們必須利用氣溫最高的夏天，在短短的一季之內完成萌芽、成長、開花、結果的完整生活史。而當年產生的種子，掉落地面之後並不會馬上發芽，而是靜靜地等待第二年春夏氣候轉暖後，才會重新演出萌芽、成長、開花、結果的完整生命歷程。落在土裡的種子，我們稱之爲「種子庫」。種子庫必須抵擋高山冰雪期的嚴苛環境，支撐到次年夏天，才能重新接受另一次的生命挑戰，年復一年，龍膽在這塊土地上闡述著淒美而堅強的生命樂章。

Steep mountains in Taiwan created various ecological environments and nurtured highly different alpine plants. Most of those plant groups originated from temperate zone. After spread and reproduced outward, some of them found suitable environments and thus survived. During long history of adaption and evolution, some of those plant groups became different from their ancestors or relations and evolved into endemic species in Taiwan. Gentiana is the most representative example. These plants prefer high attitudes and cool weather, and mostly are annual herbs. Such characteristic made them capable adapted the extreme severe weather on high mountain regions. They have to take advantage of warmest season of the year -summer- to complete the entire life cycle of shooting, growing, blooming, and fruiting. Those seeds which produced this year would not germinate right after they drop down on the surface of earth. On the contrary, silently they stay in the earth and wait until next spring and summer, when the weather became warmer and warmer they will perform the entire life cycle again. We call those hidden seeds as "Seed Bank". Seed Bank need to resists the severe snow season of mountain region and hold on to next summer, then they would have the opportunity to facing another challenge of life. Year after year, Gentiana play the harsh but beautiful music on this island.

■玉山龍膽。

攝影／蕨類研究室

豔麗身影藏不住

龍膽（*Gentiana*）參與中國人的歷史已經很久了，早在唐宋時期，便已有文獻記載龍膽在醫藥療效方面的應用，形容這種植物「葉如龍葵，味苦如膽」，龍膽也因而得名；除了藥用之外，龍膽也是著名的草本花卉。中國西南地區，龍膽的外形與花色豔麗多變、種類繁多，為全世界的分布中心，與報春花（*Primula*）齊名，並列為當地最美艷的高山植物。幸運地，台灣因為有高山，而且山頂或稜線附近常形成箭竹草坡或岩屑型草生地，這樣的環境由於光照強烈且晝夜溫差大，正是適合龍膽生長繁衍的棲息地類型，所以我們得以在台灣看到它美麗的身影。

龍膽的花冠大多形成筒狀，五個花瓣裂片間具有大而明顯的附屬裂片，所以看來像是具有十個裂片，但附屬裂片較小且先端常不規則分叉，細看就能加以區分。龍膽的花朵在經過授粉之後，尚能持續開放數天，所以天氣晴朗時，嬌小的植株上花朵齊開，喇叭狀的花朵同時綻放的景象，就像在吹奏快樂的音符一般，燦爛地舞動人心。

基本上，台灣的平地是看不到龍膽的，且也不易栽種，要觀賞龍膽最簡單的方式，就是找個風和日麗的夏天，驅車直上合歡山，你就能沿著馬路邊發現數量最多、花色最豔麗搶眼的龍膽，像是花冠藍紫色的阿里山龍膽（*G. arisanensis* Hayata），花冠艷黃色的黑斑龍膽（*G. scabrida* var. *punctulata* s. s. Ying），以及花冠藍色的台灣龍膽（*G. davidii* var. *formosana* (Hayata) T. N. Ho）。它們雖然植株嬌小，往往隱身在箭竹或草叢內，但是花朵開放時，倩影就藏不住了；龍膽的花高貴孤傲，只在晴天才完全綻放，如果天空飄來一片烏雲遮住了陽光，龍膽的花瓣很快就會合攏起來，過程僅約10秒鐘。

物種介紹

台灣龍膽 *Gentiana davidii* var. *formosana*

為本屬植物中在台灣分布最廣者，也是唯一非特有的種類，亦分布於大陸的福建及廣東一帶。線狀披針形的葉與壺狀的花冠筒是主要的區分特徵。

黑斑龍膽 *Gentiana scabrida* var. *punctulata*

大型而艷黃色的花，是它最吸引人的利器，花冠筒喉部具有黑斑是它的招牌；主要分布於海拔2,800公尺以上的裸露草生地或碎石坡，雖是一年生草本，且處於如此特殊而惡劣的環境下，但它似乎仍適應有道。

攝影／陳志雄

美麗與哀愁

台灣的龍膽屬植物共有13個種類，其中12種僅特產於台灣，明確地告訴我們它們的重要與珍貴。高貴暨美麗，時常遭天忌，這些特有種當中，半數是稀有的種類，它們稀有最主要的原因，在於它們只生長在特殊的生育環境。舉例來說，高山龍膽（*G. horaimontana* Masamune）只長在海拔3,600公尺以上的圓柏灌叢下，高雄龍膽（*G. kaohsiungensis* C. H. Chen & J. C. Wang）僅分布於中央山脈南端的裸露岩屑地，塔塔加龍膽（*G. tatakensis* Masamune）零星分布在中央山脈中南段的高海拔碎石坡，伊澤山龍膽（*G. itzershanensis* Liu & Kuo）只分布於雪山山脈的高海拔山區，厚葉龍膽（*G. tentyoensis* Masamune）與太魯閣龍膽（*G. tarokoensis* C. H. Chen & J. C. Wang）則僅生長於東部中高海拔的裸露石灰岩地形。

上述的這些特殊生育環境不僅面積有限、分布不連續而且變動性很高，對這些一年生的龍膽來說，顯然危機四伏，非常不利其傳宗接代的任務；也因為對棲地環境的選擇性狹隘，常造成族群的分布零碎化，各擁山頭互不往來，並且建立的族群小而容易受干擾。根據野外的調查經驗，生育地崩塌或地貌改

阿里山龍膽 *Gentiana arisanensis*

具有鑿形葉，開藍紫色的花，通常生長於高海拔山區的草地上或邊緣。因為是多年生草本，所以除了冰封期之外，均可在高海拔地區發現其蹤跡。

玉山龍膽 *Gentiana scabrida*

常見於高山草坡的黃色花種類，與黑斑龍膽類似，植株與花則明顯比後者小。此種類在台灣的分布甚廣，高海拔山區的開闊草地多能發現，族群穩定。

高雄龍膽 *Gentiana kaohsiungensis*

至1999年才發表的種類，形態與玉山龍膽相似，但花萼筒裂片則為線狀三角形，植株分枝常成低矮平匐狀，則與玉山龍膽明顯不同；僅分布於中央山脈南端的裸露岩屑地，族群分布侷限且稀少，須保護。

攝影／陳志雄

變，常讓稀有種類的族群飽受摧殘，即使是些微的地形變化，也可能會讓種子庫遭受嚴重的損失，受到嚴重干擾的族群，甚至隔年即消失不復見。

珍貴稀有應細心保護

經過多年調查的結果，發現稀有的龍膽大多分布局限，而且族群數量少得可憐，有些種類甚至只發現過兩、三個小族群，而且個體數量通常很少；雖然它們大多生長在國家公園的範圍內，生育棲地暫時不至於受到毀滅性的破壞，但是脆弱的族群不僅要承受嚴苛環境的考驗，還要面對棲息地經常變動的危機，一旦遭受不經意的人為干擾，依然岌岌可危。

其實在台灣的高山植物，有很多種類與龍膽相似，面臨生育地侷限與族群零碎化的問題，它們有著亮麗的生命，但也極度脆弱；這些僅存在於台灣高山的珍稀植物，不僅增添山野的生命光彩，同時也是生物基因庫的瑰寶，萬不可因為人類不智的過度利用與干擾，讓它們莫名其妙地從自然界的舞台消失。認識與了解龍膽，不僅僅是從珍稀植物的保育角度出發，整個過程更是保護高山植物資源的引子。我們對高山植物了解得越多，也才越能感受它們的珍貴與不凡。

厚葉龍膽 *Gentiana tentyoensis*

葉片厚革質，開藍色至藍紫色的花；此種類不僅族群小，而且分布零散，因為主要以裸露石灰岩地形為生育地，因生育地容易變動，使族群遭受威脅，數量稀少且瀕危。

高山龍膽 *Gentiana horaimontana*

植株呈綠黃色而小型，通常僅3~5公分高。只長在海拔3,600公尺以上的圓柏灌叢下，目前僅發現分布於玉山及馬博拉斯山兩處，個體數極稀少。

塔塔加龍膽 *Gentiana tatakensis*

葉子小而柔弱，花白色或略帶紫色，花冠筒明顯小且窄。零星分布在中央山脈中南段的高海拔碎石坡，生育地間斷且數量稀少。

攝影／陳志雄

散佈大海的星辰

離島

Acompanied Islets

◎撰文／郭城孟 ◎ Text ／ Chen-Meng Kuo

攝影／蕨類研究室

綠島。攝影／蕨類研究室

台灣位在歐亞陸版塊與菲律賓海版塊交界之處，台灣島的主體即是此兩版塊互相推擠而隆生出海水面。在此版塊交界處，火山活動旺盛，火山噴發不僅造成陸域地區的大屯火山、基隆火山等，在台灣島的四周，亦形成大大小小的火山島嶼，如北方的彭佳嶼、棉花嶼、花瓶嶼；基隆外海的基隆嶼；宜蘭的龜山島、龜卵島；台東的綠島、蘭嶼、小蘭嶼；鵝鑾鼻外海的七星岩；以及西邊的澎湖群島等都是，台灣的離島除了小琉球是珊瑚礁島之外，全為火山島。

台灣是大陸性島嶼，島上的生物受到亞洲大陸的影響極深，許多生物是由大陸遷徙來台，尤其是冰河時期台灣成為大陸生物的避難所，所以兩地生物之親緣關係密切，但是因為具有地理上的隔離，所以

Taiwan is located at the position which is between Eurasia and Philippine tectonic plate and was formed by their squeeze to each other. Volcanic activity here was very frequent and created many mountains such as volcano Ta-tung, volcano Chi-lung, and as well as created many islands around Taiwan such as Peng-chia-yu Island, Man-hua-yu Island and Hua-ping-yu Island north to Taiwan; Chi-lung-yu Island outside Chi-lung; Kue-shan Island, Kue-juan Island east to I-lan; Green Island, Lan-yu Island and Hsiao-lan-yu Island south-east of Tai-tung; Chi-hsing Island south of O-luan-pi and Peng-hu Archipelago west to Taiwan. All of these islands around Taiwan are volcanic islands except Hsiao-liu-chiu which is a coral reef island.

Because Taiwan is a continental island, the organism is affected by Asia continent deeply. Most of the organism came from Asia continent especially at The Ice Age when Taiwan was the refuge for many species. This makes the organism re-

■龜山島尾部的水池。

台灣的生物有其獨立演化之空間，因此台灣會產生與亞洲大陸不同的特有種或特有亞種。而台灣的離島相對於台灣，也一樣有類似的特性，但因距離較近，仍有交流的機會，所以大多祇是分化出不同型的族群，還不到「種」的階段。

而台灣位於東亞及東南亞植物區系的交會區，而台灣四周圍的離島，更扮演了生物遷徙中繼站的角色：亞洲區系的種類，由北方、西方進入台灣，東南亞區系則由南方的蘭嶼、綠島等，過渡到恆春半島地區，此從島嶼上的植物分布類型即可得證，可惜北部及西部的島嶼，以及東南部的綠島，多受人類活動干擾，植被受到嚴重破壞，特色較不明顯，蘭嶼和龜山島則尚保留了相當多的自然植被，是研究生物地理的絕佳材料。

lationship between Taiwan and Asia continent closely. However, because of the geography separation, the organism on Taiwan has chance to evolve independently and then result in many endemic species and endemic subspecies here. The offshore islands around Taiwan have the same characteristic as Taiwan. But many of their species can not reach the level of species, they only differentiate to different population as a result of that these islands are close enough to Taiwan proper.

In addition to that Taiwan is located at the interface between Sino-Japanese, and S. E. Asiatic floristic zones. The islands around Taiwan especially function as the medium when the organism entered Taiwan. For example, the Sino-Japanese species came from north and west; S. E. Asiatic species came from south through Lan-yu Island to Heng-chung Peninsula. This phenomenon can be proved by the distribution of flora on those islands. However, it is a pity that the islands at north, west and Green Island which is located at south-east are disturbed and destroyed severely by human and make their vegetation characteristic less obvious. Fortunately Lan-yu Island and Kue-shan Island still conserve lots of natural vegetation and could be great objects of biogeography research.

攝影／蕨類研究室

海外仙山

離島植被

Vegetation of Island

◎撰文／楊遠波

◎ Text ／ Yuen-Po Yang

彭佳嶼的植被

彭佳嶼位於基隆東北方約55公里處，是一年輕的火山島，島略呈梯形，地勢大致東高西低，東、南及北面皆為斷崖，西側成緩坡，海岸多礁石，環島約4,300公尺，面積約114公頃。

全島植被以白茅（*Imperata cylindrical* (L.) P. Beauv.var.*major* (Nees) C.E.Hubb.）及芒草為優勢所形成的草原為主。地勢較高且衝風處以白茅為主，偶見細葉饅頭果（*Glochidion rubrum* Blume）及朝鮮紫珠（*Callicarpa japonica* Thunb.var. *luxurians* Rehd.）混生其間，高度多不及腰。地勢低且土壤發育較佳處，則以芒草（*Miscanthus sinensis* Anders.）為優勢，可達一人高，其根部常有野菰（*Aeginetia indica* L.）寄生。地勢較凹處或岩石的背風面常形成矮灌叢，組成物種以島榕（*Ficus virgata* Reinw.&Bl.）、小葉桑（*Morus australis* Poir.）、青苧麻（*Boehmeria nivea* (L.)Gaudich. var.*tenacissima* (Gaudich.)Miq.）、朝鮮紫珠及蘭嶼樹杞（*Ardisia elliptica* Thumb.）為主，常聚集成團狀，高度依風勢而定，多不過2米，或因風及鹽害之故，樹多分枝且頂端常呈枯枝狀。沿岸地區僅西側有植被覆蓋，以白水木（*Tournefortia argentea* L. f.）、草海桐（*Scaevola sericea* Forster f.）及苦林盤（*Clerodendrum inerme* (L.) Gaertn.）為主的灌叢，呈帶狀分布於海邊的礁岩上。西南側有一陡坡，此地區較為陰暗潮濕，常見全緣貫眾蕨（*Polystichum falcatum* (L.f.) Diels）、鐵線蕨

■彭佳嶼全島植被以白茅及芒草為優勢所形成的草原。

攝影／陳添財

■彭佳嶼位於基隆東北方約55公里處，是一年輕的火山島，島略呈梯形，地勢大致東高西低，東、南及北面皆為斷崖，西側成緩坡，海岸多礁石。

攝影／陳添財

（*Adiantum capillus-veneris* L.）、闊片烏蕨（*Sphenomeris biflora* (Kaulf.) Tagawa）、鴨趾草（*Commelina communis* L.）及脈耳草（*Hedyotis strigulosa* Bartl. *ex* DC. var.*parvifolia* (Hook.&Arn.)Yamazaki）等較喜陰濕的種類，生長於岩縫或礫石中。

龜山島的植被

龜山島位於宜蘭外海，是火山形成的島嶼，全島面積2.7平方公里。冬季受到東北季風的影響，氣溫降低，其強風使得林木矮化、緊密且不分層次。此外，島上斷崖處生長著天然蒲葵（*Livistona chinensis* (Jacq.)R. Brown var. *subglobosa* (Martius.) Beccari）族群。

可能因為過去島上居民以海為生，耕作面積不大，島上植被受干擾小。大致上，海拔260公尺以上仍是原生的天然林，以下則是廢耕後形成的次生植被。基本上此地的植被可分成海岸植被、次生植被和山地植被。

■從龜山島山上，下望尾部的礫石灘地。 攝影／陳添財

海岸植被分布於靠近海邊地區，組成的植物多能適應海邊強風及鹽沫等環境，主要植物有海檬果（*Cerbera manghas* L.）、歐蔓（*Tylophora ovata* (Lindl.)Hook. ex Steud.）、石板菜（*Sedum formosanum* N.E. Br.）、小毛蕨（*Cylosorus acuminata* (Houtt.)Nakai ex H. Ito.）、越橘葉蔓榕（*Ficus vaccinioides* Hemsl.）、海埔姜（*Vitex rotundifolia* L.f.）、過江藤（*Phyla nodiflora* (L.) Greene）、茅毛珍珠菜（*Lysimachia mauritiana* Lam.）、天蓬草舅（*Wedelia prostrata* (Hook. & Arn.) Hemsl. var. *prostrata*）等等。另可見印度鞭藤（*Flagellaria indica* L.）、毛柿（*Diospyros philippensis* (Desr.) Gurke.）、黃心柿（*Diospyros maritima* Blume）和琉球澤蘭（*Eupatorium luchuense* Nakai）等。

次生植被可分成草本群落和次生林群落。草本群落在龜尾平地從海岸地區到森林邊緣，由於接近海濱，有許多植物和海岸植被相同，如石板菜、過江藤等等。可能因長期受到以往居民的影響，植株多為小草本及蔓性植物，例如鐵牛入石（*Cocculus orbiculatus*(Linn.) DC.）、馬鞭草（*Verbena officinalis* L.）、磚子苗（*Cyperus cyperiodes* (L.) O. Kuntze）等。此外，在較高地區可見五節芒（*Miscanthus floridulus* (Labill) Warb. ex Schum. & Laut.）、芒萁（*Dicranopteris linearis* (Barm. f.) Underw.）、金絲桃（*Hypericam monogynum* L.）形成的小面積草生群落。次生林分布於島嶼西、北部海拔260公尺以下的闊葉林，是經過人為干擾後恢復的天然林，主要樹種有江某（*Schefflera octophylla* (Lour.) Harms.）、野桐（*Mallotus japonicus* (Thunb.) Muell. - Arg.）、山黃麻（*Trema orientalis* (L.) Bl.）、山香圓（*Turpinia formosana* Nakai）、筆筒樹（*Cyathea lepifera* (Hook.) Copel.）、相思樹（*Acacia confusa* Merr.）、桂竹（*Phyllostachys makinoi* Hayata）等，和台灣北部低海拔山區的植物種類相似程度極高。

山地植被在海拔260公尺以上，直到最高峰。此地帶受強烈東北季風吹拂，植株多矮小而呈灌木化的現象，復因溫度降低，出現暖溫帶的植物，主要樹種有金平氏冬青（*Ilex maximowicziana* (Thumb.) Makino）、凹葉柃木（*Eurya emarginata* (Thumb.) Makino）、厚葉衛矛（*Euonymus carnosus* Hemsl.）、米飯花（*Vaccinium bracteatum* Thunb.）、馬醉木（*Pieris taiwanensis* Hayata）、粗糠柴（*Mallotus philippensis* (Lam.)Muell.-Arg）、大明橘（*Myrsine sequinii* H.Levl.）等等。

澎湖的植被

澎湖本島及其離島共計有大小64座。其中澎湖本島、白沙島和西嶼三個主要島嶼即佔澎湖群島全部面積的百分之八十五。這裡的氣候相當乾燥、風多又強是最大特色。由於開發甚早，植被遭受長

■西嶼（澎湖群島之一小島）島上的黃花酢漿草，為草生地上之一主要種類。 攝影／施炳霖

期且嚴重的人為干擾，復因氣候的原因，所有島嶼幾不見天然森林的存在(是否有原生的森林仍待考證)，除了人工栽植的植被和少數天然灌叢外，此地的植被是以草本為主。

目前所知，澎湖本島大約有282種植物。木本植物約15種且大部份零星分布島上。常見的有天人菊（*Gaillardia pulchella* Foug.）、木麻黃（*Casuarina equisetifolia* L.）、銀合歡（*Leucaena leucocephala* (Lam.) de Wit）、仙人掌（*Opuntia dillenii* (Ker) Haw.）和欖仁（*Terminalia catappa* L.）等，都是栽植或馴化的植物。木麻黃和欖仁是引進的造林樹種。銀合歡、天人菊和仙人掌是馴化的種類，常成片生長形成各個群落。島上可見大片的草海桐群落，是原生的種類被大量栽植作為防風用途。此外，島上有少數零星天然生長的林投（*Pandanus odoratissimus* L.f.）和苦林盤各個灌叢。

島上的天然草生植被最主要分布在海邊。海邊岸上的馬尼拉芝（*Zoysia matrella* (L.) Merr.）群落面積最大，其他尚有以豆科、禾本科或莎草科為主佔小面積的群落。在海水裡的是卵葉鹽藻（*Halophila ovalis* (R. Br.) Hook. f.）和單脈二藥藻（*Halodule uninervis* (Forsk.) Aschers.）的植物群落。

■西嶼島上草生植被中的石莼蓉。　攝影／施炳霖

小琉球的植被

小琉球位於屏東縣東港鎮西南方約12公里的海上，面積約6.8平方公里，是由珊瑚礁岩所形成的小島。居民以漁業及觀光業為主，全島除住宅週邊偶見的菜園及果樹外，幾無農業的耕作，由荒地所形成的次生植被，是島上面積最大的植被類型。

全島的植被約可區分為人工及天然植被兩大類。人工植被以刺竹（*Bambusa stenostachya* Hackel）、相思樹（*Acacia confusa* Merr.）及鳳凰木（*Delonix regia* (Bojer ex Hook.) Raf.）呈較大面積的分布，其餘果樹、行道樹及庭園植物等，多沿村落或公路兩側分布，並無所謂的植被類型可言。陸生的

上■小琉球隆起的珊瑚礁或臨海坡面上，有較類似於海岸林的植被類型。
左一■小琉球的村落附近的人工植被以刺竹、相思樹及鳳凰木的面積較大。
左二■小琉球島上道路旁廢墾地上常見的牧地狼尾草成片生長。
左三■小琉球全島的次生林多呈低矮及密生狀，高度多不超過三米。

攝影／陳添財

天然植被大略可分為海岸及次生植被兩類。在美人洞、大寮等地的珊瑚礁岩及海蝕平臺上，常有灌叢狀的植物分布，以苦林盤、林投、雙花蟛蜞菊（*Wedelia biflora* (L.) DC.）及肥豬豆（*Canavalia lineata* (Thunb. ex Murray) DC.）為主要組成物種。海崖或衝風處的草生地上，以升馬唐（*Digitaria ciliaris* (Retz.) Kotler）、土丁桂（*Evolvulus alsinoides* (L.) L.）、白花草（*Leucas chinensis* (Retz.) R. Br.）及無根草（*Cassytha filiformis* L.）為常見，偶有呈匍伏狀的紅仔珠（*Breynia vitis-idaea* (Burm. f.) C.E. Fischer）、烏柑仔（*Severinia buxifolia* (Poir.) Tenore）及滿福木（*Carmona retusa* (Vahl) Masam.）與其混生。隆起的珊瑚礁或臨海坡面上，有較類似於海岸林的植被類型；臨海的礁岩上以林投、黃槿（*Hibiscus tiliaceus* L.）、葛塔德木（*Guettarda speciosa* L.）、榕樹（*Ficus microcarpa* L. f.）及山豬枷（*Ficus tinctoria* Forst. f.）為常見，山坡上以大葉雀榕（*Ficus caulocarpa* (Miq.) Miq.）、雀榕（*Ficus superba* (Miq.) Miq. var. *japonica* Miq.）、茄苳（*Bischofia javainca* Blume）及榕樹為優勢，林下則有姑婆芋（*Alocasia odora* (Lodd.) Spach.）、月桃（*Alpinia zerumbet* (Persoon) B. L. Burtt & R. M. Smith.）、竹葉草（*Oplismenus compositus* (L.) P. Beauv.）、青苧麻等灌木或草本植物，藤本植物以風藤（*Piper kadsura* (Choisy) Ohwi）、拎樹藤（*Epipremnum pinnatum* (L.) Engl. ex Engl. & Kraus.）及三角葉西番蓮（*Passiflora suberosa* L.）等較常見。次生植被多由廢墾地演替而來，有草生地、灌草叢及次生林等類型之分，幾乎遍佈全島。草生地及灌草叢因季節的更替，可能有不同的組成物種，常見的有山地豆（*Alysicarpus vaginalis* (L.) DC. var. *vaginalis*）、大花咸豐草（*Bidens pilosa* L. var. *radiata* Sch. Bip.）、賽葵（*Malvastrum coromandelianum* (L.) Garcke）、長柄菊（*Tridax procumbens* L.）、狗牙根（*Cynodon dactylon* (L.) Per.）、牧地狼尾草（*Pennisetum polystachion* (L.) Schult.）、白茅、甜根子草（*Saccharum spontaneum* L.）、五節芒、銀合歡（*Leucaena latisiliqua* (L.) Gillis）、馬纓丹（*Lantana camara* L.）、美洲含羞草（*Mimosa diplotricha* C. Wright ex Sauvalle）、細葉饅頭果、白飯樹（*Flueggea suffruticosa* (Pallas) Baillon）及紅仔珠等。全島的次生林多呈低矮及密生狀，高度多不超過三米，主要組成物種有黃荊（*Vitex negundo* L.）、銀合歡、血桐（*Macaranga tanarius* (L.) Muell.-Arg.）、構樹（*Broussonetia papyrifera* (L.) L' H'erit. ex Vent.）、恆春厚殼樹（*Ehretia resinosa* Hance）、山柚（*Champereia manillana* (Bl.) Merr.）、相思樹及月橘（*Murraya paniculata* (L.) Jack.）等，常見的林下草本有鱗蓋鳳尾蕨（*Pteris vittata* L.）、密毛毛蕨（*Cyclosorus parasitica* (L.) Farw.）、海金沙（*Lygodium japonicum* (Thunb.) Sw.）及竹葉草等。常出現於次生植被的藤本植物有盒果藤（*Operculina turpethum* (L.) S. Manso.）、扛香藤（*Mallotus repandus* (Willd.) Muell.-Arg.）、蝶豆（*Clitoria ternatea* L.）、槭葉牽牛（*Ipomoea cairica* (L.) Sweet.）、蔓澤蘭（*Mikania cordata* (Burm. f.) B.L. Rob.）、雞屎藤（*Paederia foetida* L.）及紅花野牽牛（*Ipomoea triloba* L.）等。

■蘭嶼島上的植被依組成、形相和分布地帶可區分成海濱植被和山地植被。海濱植被從海水高潮線上的海邊向內陸延伸到山腳和斷崖基部，整個在島上呈一狹窄的環帶圍繞島中央台地山區。

攝影／廖俊奎

蘭嶼、綠島及小蘭嶼植被

蘭嶼、小蘭嶼與綠島位於台灣本島的東南方，各有面積約45.7、1.6和15.4平方公里；前兩島嶼相距約5公里，兩者與綠島相距約60公里。因為這三個島距離近，島上的植物種類重疊性高，尤其是小蘭嶼的植物全數在蘭嶼島上可見。大致上，三個島上的植物多數可見於台灣本島，其次菲律賓，最後是琉球，這現象與三島和三地區的距離相關。

氣候上，蘭嶼、小蘭嶼和綠島的大致相仿，在植物組成相近的情況下，三者的植被形相也相似。由於蘭嶼的山地海拔較其他兩者高，面積也大，島上的植被因而較為複雜。因此，將以蘭嶼的植被作為其他二者的代表，隨後將另述另二島的現況。

蘭嶼氣候經年高溫、高濕、多雨又多風，可說很接近熱帶雨林氣候的性質。島上大部分區域為山地，僅少部分平地。山脈大致分南北二部分，二者交會於島的東南部呈鞍部形。簡言之，山地區雖然最高處海拔548公尺，但山坡地及河谷均陡峭，整體呈一切割台地的地形。島上的植被依組成、形相和分布地帶可區分成海濱植被和山地植被。

海濱植被從海水高潮線上的海邊向內陸延伸到山腳和斷崖基部，整個在島上呈一狹窄的環帶圍繞島中央台地山區。海濱植被大致以與海水的距離由近至遠可分草本植物群落、低矮灌木形成的灌木群落、林投群落及較高大的海岸林群落。草本植物群落中常見的有高麗芝（*Zoysia tenuifolia* Willd. ex Trin.）及乾溝飄拂草（*Fimbristylis cymosa* R. Br.）。灌木群落中有苦林盤、海埔姜、白水木等等。林投群落是以林投為主。海岸林群落中佔多數的有棋盤腳樹（*Barringtonia asiatica* (L.) Kurz.）、皮孫木（*Pisonia umbellifera* (Forst.) Seem.）和蓮葉桐（*Hernandia nymphiifolia* (Presl) Kubitzki.）等等。

山地植被從山腳及斷崖基部往內陸山地分布。植被的組成大多數與海濱植被不同，僅在臨海斷崖壁上形成一過渡群落，是由山地植被的成分及海濱植被的成分混生，但以蔓藤類植物為主，如雙花蟛蜞菊、鵝鸞鼻蔓榕（*Ficus microcarpa* L.f. var *oluangpiensis* Liao.）和山豬枷等等。

山地植被以形相和組成可分草本植物群落、灌叢植物群落、濕生群落和森林群落。

山地草本植物群落以芒草為主，分布島上北部和東部的陡峭坡面。山地灌叢群落長在島上少數的主稜線上，因為強風的吹襲，木本植物多不及2公尺高，以山林投（*Freycinetia formosana* Hemsl.）為主要成分。山地濕生植物群落面積不大，僅在島上的天池四周，主要的成分是草本的長箭葉蓼（*Polygonum hastatosagittatum* Makino.）和畦畔莎草（*Cyperus haspan* L.）。森林群落分布在分水嶺、山峰主稜線、山腹或溪谷的河階土壤深厚處。森林中以白榕、蘭嶼福木（*Garcinia linii* C.E. Chang）、茄苳、小葉樹杞（*Ardisia quinquegona* Blume）、貝木（*Timonius arboreus* Elmer）、銹葉野牡丹（*Astronia ferruginea* Elmer）為主要的成分。此外，尚有由番龍眼構成的純林，位於低海拔地區。據謂，此純林可能為人工林。

除了上述的天然植被外，在蘭嶼環島的平地和平緩的山丘常有人工放牧的草原，以褐色狗尾草（*Setaria pallide-fusca* (Schumach.) Stapt & C.E. Hubb.）、短軸秀竹（*Microstegium glaberrimum* (Honda) Koidz.）、亨利馬唐（*Digitaria henryi* Rendle）、圓果雀稗（*Paspalum orbiculare* G. Forst.）和白茅為主要成分。

小蘭嶼的植被於數年前因受火災及島上放羊的影響，多為草生植被，僅在島上低坳山溝地區有森林植被的存在，在海邊及山坡坡面有灌叢的生長，其成分主要為林投，間亦混生四脈麻（*Leucosyke quadrinervia* C. Robinson）。草生植被主要為白茅、八丈芒（*Miscanthus sinensis* Anders.）和扁莎（*Pycreus* spp.）屬的植物。

綠島由於開發較早，島上原生植被多被濫伐幾無所存，在山溝地區及過山步道，乃可以見到一些原生植群，植物種類則介於台灣本島和蘭嶼之間，乃是以一些蕁麻科、桃金孃科、大戟科、楝科及山欖科等熱帶區系植物為主。

■小天池位於開闊的寬稜之上。 攝影／廖俊奎

■山地濕生植物群落面積不大，僅在島上的天池四周。可分為草本植物帶與木本植物帶。 攝影／廖俊奎

■小蘭嶼的稜線主要為草生植被，較低窪或避風處亦可見灌叢群落。 攝影／廖俊奎

■綠島的原生植被多數已遭受破壞，次生林為島上主要的植被。 攝影／陳添財

海上遺珠

蘭嶼

Lanyu

◎撰文／楊宗愈
◎Text ／T.Y. Aleck Yang

蘭嶼，台灣東南部一個遺世獨立的島嶼，然而因爲地理位置剛好在台灣、琉球和菲律賓之間，所以造成島上植物種類異常地豐富，據估計約有800種以上，其組成份較偏屬於熱帶，若就木本植物而言則與菲律賓植群較爲相似，若依植物總數則和台灣植物區系關係較近，所以明顯是一「交匯帶」或「過渡橋樑」的角色。爲了方便介紹，本文將蘭嶼分成海濱植物群落、草原植物群落、灌叢植物群落、濕生植物群落、森林植物群落及作物植物群落等六大區域說明。

Lanyu, an isolated island from a lost age off the east coast of Taiwan, has become extraordinarily rich in plant species because of its favorable geographical position among Taiwan, Liuchiu and the Philippines. It is estimated that there over 800 species on the island, most of which are tropical. In terms of tree species, there are greater similarities to the species similar withn the Philippines; for plants in general, the species are more related to those seen on Taiwan floristic. From this we can see that this island clearly plays a role as a crossroads or bridge. For the sake of simplicity, this paper will divide the plant life on Lanyu into coast vegetation, grassland vegetation, shrubs, swamp vegetation, forests and crop plants.

■蘭嶼。

攝影／李明宜

■台灣蝴蝶蘭。 攝影／楊宗愈

蘭花的島嶼

「蘭嶼」(Orchid Island) 蘭花的島嶼！多麼好聽的名字，這個位在台灣東南方的小島，曾經被叫作「Botel Tobago」、「紅頭嶼」，而住在島上的住民—達悟族（雅美人）則是叫他們生長的島嶼為「do Irala」，有「靠內陸的、離海較遠的、北邊」等意思，外地人則稱該島為「pongso no tao」意思是「人的島嶼」。而「蘭嶼」這個名字，也真得是因為蘭花的關係；原來在民國三十六年（1947）在一項國際的花卉展覽中，產于島上潔白的台灣蝴蝶蘭獲得該項展覽的冠軍，而「紅頭嶼」（在當時的稱呼）即被建議更改為「蘭嶼」。

有關蘭嶼的植物調查，最早的紀錄是在1896年日本人矢野氏，曾前往蘭嶼採集過；之後陸陸續續地又有許多日本學者（包括動物、植物、地質及人類學域等）、歐洲學者及光復後的台灣學者，均曾前往蘭嶼進行長短不等調查及採集。整合前人報告、資料、經驗等等，大致也將蘭嶼的植被分成六大區：海濱植物群落、草原植物群落、灌叢植物群落、濕生植物群落、森林植物群落及作物植物群落。當然這種區分法只是簡單的區別，還有許多是介於其間或是特殊的景觀，將再為文介紹說明。

海濱植物群落

由於蘭嶼是個島嶼，四面環海，加上地形特殊，所以對海濱植物群落（Coast vegetation）的定義是指那些沿海分布，其寬度（與海岸線垂直）由20公尺到500公尺不等其間的砂灘或珊瑚礁地區的植被，又由於本島與位置特殊，介於熱帶與亞熱帶植物區及親潮、黑潮二潮流間，故植物種類、層次都非常豐富，從草本植物、灌木、灌叢到大喬木，在本群落中均可以找到。由於生育環境及植被類群的差異，又再將本群落細分成珊瑚礁植被、砂地植被、熱帶海岸林植被及高位珊瑚礁植被。

珊瑚礁植被可以說是面海的第一線，所以植物多數為低矮的「草本植物」，例如：乾溝飄拂草（*Fimbristylis cymosa* R. Br.），高麗芝（*Zoysia tenuifoloia* Willd. ex Trin.），耳草屬植物（包括雙花耳草*Hedyotis biflora* (L.) Lam. 及脈耳草*H. strigulosa* Bartl. *ex* DC. var. *parvifolia* (Hook. & Arn.) Yamazaki）、安旱草（*Philoxerus wrightii* Hook. f.）等及「小灌木」，例如：水芫花（*Pemphis acidula* J.R. & G. Forst.）、蘄艾（*Crossostephium chinense* (L.) Makino）、南嶺蕘花（*Wikstroemia indica* (L.) C.A. Mey.）、白水木（*Tournefortia argentea* L. f.）等，或是匍匐性的「草質莖藤本」，例如：雙花蟛蜞菊（*Wedelia biflora* (L.) DC.）、三葉木藍（*Indigofera trifoliate* L.）、爬森藤（*Parsonia* *laevigata* (Moon) Alston）等及「木質莖藤本」，例如：馬鞍藤（*Ipomoea pes-caprae* (L.) R. Br. ssp. *brasiliensis* (L.) Ooststt.）、海埔姜（*Vitex rotundifolia* L. f.）、苦林盤（*Clerodendroum inerme* (L.) Gaertn.）等；當然這些植物都必須能耐鹽（包括海水及海霧）及耐強風，所以我們可以看到上述這些植物的莖葉，多數都是很厚或者就是表面密佈著毛茸。離海水較遠處則可以生長一些較直立的種類，例如蘭嶼小鞘蕊花（*Coleus formosanus* Hayata）、茅毛珍珠菜（*Lysimachia mauritiana* Lam.）、蘭嶼木耳菜（*Gynura elliptica* Yabe & Hayata）、蘭嶼山柑（*Capparis lanceolaris* DC.）、白木蘇花（*Dendrolobium umbellatum* (L.) Benth.）、毛苦參（*Sophora tomentosa* L.）、臭娘子（*Premna serratifolia* L.）、林投（*Pandanus odoratissimus* L. f.）等也都可以生長。整體上說來，在珊瑚礁地區生長的植物主要都是那些低矮且抓地力強、耐鹽性強、耐強風吹、匍匐性或是具有支柱根的種類，共約在100種上下。

■蘭嶼羅漢松。　　攝影／楊宗愈

蘭嶼典型的砂地僅在東部的東清村，明確地說約在環島公路12公里處及17~18公里間的一小段砂地；該等處的砂地植被以白水木、南嶺蕘花等及邊緣的林投、黃槿（*Hibiscus tiliaceus* L.）、檄樹（*Morinda citrifolia* L.）及臭娘子為主要灌叢或小喬木，其他長在砂地上的則多為匍匐性灌木或深根性的草本植物，例如：海埔姜、雙花蟛蜞菊、濱豇豆（*Vigna marina* (Burm.) Merr.）、濱大戟（*Chamaesyce atoto* (Forst. f.) Croizat）、牡蒿（*Artemisia japonica* Thunb.）蘭嶼白脈根（*Lotus australis* Andr.）、蘭嶼木耳菜等。由于面積不大，種類並不多，約在50種左右。

在蘭嶼，由於達悟人的信仰關係，位在珊瑚礁邊緣的熱帶海岸林植被區往往是保留最好的一片，因為這些地方都有當地人最畏懼的植物之一，那就是當地話叫做「kamanlalazoan」或「toba」的「魔鬼樹」，而這些地區也就是當地人主要的墓地之一。雖然「熱帶海岸林」有嚴格的定義，例如果實必須可以海飄，且登陸後立即可以發芽生根等等，然而有些比較靠近內陸的種類可能並不完全符合，但為了完整說明，仍將它們列入本植被區。又熱帶海岸林植被常有一個現象，即林下草本植物稀少，整個林區主要都是喬木及灌木；在蘭嶼也是一樣，除了前面提過的「魔鬼樹」，也就是玉蕊科的棋盤腳樹（*Barrintonia asiatica* (L.) Kurz.）外，還有欖仁（*Terminalia catappa* L.）、蓮葉桐（*Hernandia nymphiifolia* (Presl) Kubitzki）、大葉樹蘭（*Aglaia elliptifolia* Merr.）、皮孫木（*Pisonia umbellifera* (Forst.) Seem.）、青脆枝（*Nothapodytes nimmoniana* (Graham) Mablerley）、大葉雀榕（*Ficus caulocarpa* (Miq.) Miq.）、大冇榕（*F. septica* Burm. f.）、厚殼樹（*Ehretia acuminata* R. Br.）、黃心柿（*Diospyros maritime* Blume）、大葉山欖（*Palaquium formosanum* Hayata）等等，連同蔓藤性植物及附近的草本植物都算進去，本植被區大約有100種左右。

在珊瑚礁邊緣的隆起或較近內陸的隆起珊瑚礁，也就是本文所謂的高位珊瑚礁植被；有的珊瑚礁並不高約只有1~2公尺，而有些則可隆起至80公尺或更高。由於本植被區多數並非直接受到強烈海風的吹襲，且遮蔽也較多，所以植物類型與種類都比較豐富；在較靠近內陸的地區，有些小喬木或大喬木直接長在隆起的珊瑚礁上，這些種類憑藉著「立足點」較高，一下子就比其他種類向上高出數公尺或數十公尺嘍。看似平靜的生態區，其間也是隱藏了許多激烈殘酷得競爭。在本植被區，喬木可以麵包樹（*Artocarpus incisus* (Thunb.) L. f.）、番龍眼（*Pometia pinnata* Forst.）、刺桐（*Erythrina variegata* L.）、白肉榕（島榕）（*Ficus virgata* Reinw. ex Blume）、豬母乳（*F. fistulosa* Reinw. ex Blume）、厚葉榕（*F. microcarpa* L. f. var. *crassifolia* (W.C. Shieh) J.C. Liao）、大葉山欖、蘭嶼山欖（*Planchonella duclitan* (Blanco) Bakh. f. & Kosterm）、皮孫木、蘭嶼蘋婆（*Sterculia cerramica* R. Br.）等較多，其他如蘭嶼羅漢松（*Podocarpus costalis* Presl）、厚殼樹、蘭嶼厚殼樹（*Ehretia philippinensis* A. DC.）、落尾麻（*Pipturus arborescens* (Link) C. Robinson）、鐵色（*Drypetes littoralis* (C.B. Rob.) Merr.）、長果月橘（*Murraya paniculata* (L.) Jack. var. *omphalocarpa* (Hayata) Swingle）、對葉榕（*Ficus cumingii* Miq. var. *terminalifolia* (Elm.) Sata）、大冇榕、紅頭鐵莧（*Acalypha kotoensis* Hayata）、假鐵莧（*Claoxylon brachyandrum* Pax & Hoffm.）、蘭嶼土沈香（*Excoecaria kawakamii* Hayata）、大葉樹蘭、蘭嶼樹蘭（*Aglaia chittagonga* Miq.）、象牙柿（*Diospyro ferrea* (Willd.) Bakhuizen）、蘭嶼柿（*D. kotoensis* Yamazaki）、葛塔德木（*Guettarda speciosa* L.）等灌木也參差其中，而在高位珊瑚礁最大特色，即地

■蘭嶼白脈根。　攝影／楊宗愈

面匍匐或向上攀爬藤蔓性的灌木、草本植物特別多，例如：錐頭麻（*Poikilospermum acuminatum* (Trecul) Merr.）、港口馬兜鈴（*Aristolochia zollingeriana* Miq.）、刺裸實（*Maytenus diversifolia* (A. Gray) D. Hou）、鵝鑾鼻蔓榕（*Ficus pedunculosa* Miq. var. *mearnsii* (Merr.) Corner）、山豬枷（*F. tinctoria* Forst. f.）、小葉黃鱔藤（*Berchemia lineata* (L.) DC.）、菲律賓胡椒（*Piper philippinum* Miq.）、蘭嶼風藤（*P. arborescens* Roxb.）、紅葉藤（*Rourea minor* (Gaertner) Leenhouts）、鵝掌藤（*Schefflera odorata* (Blanco) Merr. & Rolfe）、毬蘭（*Hoya carnosa* (L. f.) R. Br.）、栫樹藤（*Epipremnum pinnatum* (L.) Engl. ex Engl. & Kraus）等，再加上礁石上的多種草本植物，例如海岸星蕨（*Microsorium scolopendrium* (Burm.) Copel.）、抱樹石葦（*Pyrrosia adnascens* (Sw.) Ching）、台灣佛甲草（*Sedum formosanum* N.E. Br.）、蘭嶼秋海棠（*Begonia fenicis* Merr.）、日本前胡（*Peucedanum japonicum* Thunb.）、細葉假黃鵪菜（*Crepidiastrum lanceolatum* (Houtt.) Nakai）、粗莖麝香百合（*Lilium longiflorum* Thunb. var. *scabrum* Masam.）、闊葉麥門冬（*Liriope platyphylla* F.T. Wang & T. Tang）。若就整個海濱植物群落而言，本植被區因地形的複雜及生育地的多樣，所以種類最為豐富，包括草本植物、灌木、喬木及藤蔓性植物估計至少有250種以上。

草原植物群落

可能由於當地人是火燒山的方法來取得許多的耕地，而當這些耕地廢棄後，往往形成以芒草等禾本科植物為主的草原植物社會。而在蘭嶼島，本種植物群落比較明顯或是面積較大的，主要是在島的北面、東面及南面臨海之山崖、陡峭坡面或近稜線處。

雖然說草原植物群落（Grassland vegetation）是以禾本科植物為主，但在其間或是邊緣仍有許多灌木或其他科的草本植物生長，最為明顯的就是對葉榕、大冇榕、蘭嶼馬蹄花（*Tabernaemontana subglobosa* Merr.）、草野氏冬青（蘭嶼冬青）（*Ilex kusanoi* Hayata）、山菊（*Farfugium japonicum* (L.) Kitam.）、白木蘇花、台灣佛甲草、華南大戟（*Chamaesyce vachellii* (Hook. & Arn.) Hurusawa）、細葉油柑（*Phyllanthus virgatus* Forst. f.）、蘭嶼小鞘蕊花、一枝香（*Vernonia cinerea* (L.) Less.）、粗莖麝香百合、林投、台灣月桃（*Alpinia formosana* K. Schumann）等，而草原的主角除了是禾本科的八丈

代表物種

台灣蝴蝶蘭 *Phalaenopsis aphrodite* Reichb.
蘭科

多年生附生的蘭花，莖非常的短，被肉質橢圓形或長橢圓形的葉子所包裹著；花朵純白色，直徑約有6公分長，唇瓣尖端有2枚尾狀物。在台灣僅原生於恆春半島和蘭嶼，現在非常稀少。

蕲艾 *Crossostephium chinense* (L.) Makino
菊科

半匍匐性的低矮灌木，葉片銀白兩面被毛，三（至五）出羽狀裂片，搓揉有香味；頭狀花黃色，排成總狀花序，直徑約0.5公分。主要分布在海邊的珊瑚礁區，常常被人們挖掘栽植，做為盆景或草藥。

攝影／楊宗愈

芒（*Miscanthus sinensis* Anders var. *condensatus* (Hack.) Makino）以外，尚有水蔗草（*Apluda mutica* L.）、毛臂形草（*Brachiaria villosa* (Lam.) A. Camus）、升馬唐（*Digitaria ciliaris* (Retz.) Koeler）、兩耳草（*Paspalum conjugatum* Bergius）、鴨母草（*P. scrobiculatum* L.）、囊穎草（*Sacciolepis indica* (L.) Chase）、莠狗尾草（*Setaria geniculata* P. Beauv.）、御谷（*S. glauca* (L.) P. Beauv.）、鼠尾粟（*Sporobolus indicus* (L.) R. Br. var. *major* (Buse) Baaijens）等，及卵形飄拂草（*Fimbriatylis ovata* (Burm. f.) J. Kern）莎草科植物。由於生育環境的關係，本植物群落的種類並不多，僅約有50種，但一般說來，每一種的族群數量都還不小。

灌叢植物群落

蘭嶼島上最明顯且較容易抵達的灌叢植物群落（Shrub vegetation），主要是在西北隅燈塔及「小天池」等主稜線處。植物型由於強風的因素，所以多為3公尺以下的灌木或低矮的灌叢，而若是生長在背風面，同一種植物則可以長到6公尺高；概括地說，本植物群落是以山林投及一些蕁麻科、樟科、大戟科、桃金孃科、紫金牛科、茜草科等植物為主。

雖受風力的影響無法長得很高，但仍略呈小喬木樣的有大葉山欖、金新木薑子（*Neolitsea sericea* (Blume) Koidz. var. *aurata* (Hayata) Hatusima）、紅楠（*Machilus thunbergii* Sieb. & Zucc.）、澀葉榕（*Ficus irisana* Elm.）、大冇榕、蘭嶼樹蘭、賽赤楠（*Acmena acuminatissima* (Blume) Merr. & Perry）、蘭嶼柹、呂宋毛蕊木（*Gomphandra luzoniensis* (Merr.) Merr.）、台東漆樹（*Semecarpus gigantifolia* Vidal）、蘭嶼紫金牛（*Ardisia elliptica* Thunb.）、小葉樹杞（*A. quinquegona* Blume）、山欖（樹青）（*Planconella obovata* (R. Br.) Pierre）、菲律賓朴樹（*Celtis philippensis* Blanco）、山黃梔（*Gardenia jasminoides* Ellis）、呂宋水錦樹（*Wendlandia luzoniensis* DC.）、厚殼樹、蘭嶼虎皮楠（*Daphniphyllum glacescens* Blume var. *lanyuense* T. C. Huang）等；較常見的灌木或灌叢則有蔓榕（*Ficus pedunculosa* Miq.）、菱葉濱榕（*F. tannoensis* Hayata form. *rhombifolia* Hayata）、四脈麻（*Leucosyke quadrinervia* C. Robinson）、凹葉柃木（*Eurya emarginata* (Thunb.) Makino）、細葉饅頭果（*Glochidion rubrum* Blume）、披針葉饅頭果（*G. zeylanicum* (Gaertn.) A. Juss. var. *lanceolatum* (Hayata) M.J. Deng & J.C. Wang）、蘭嶼土沈香、

水芫花 *Pemphis acidula* J.R. & G.Forst.
千屈菜科

多數呈現匍匐蔓藤狀且有許多分枝的小型灌木；單葉、對生，狹長倒卵形，肉質，沒有葉柄；花朵白色或是略帶粉紅，花萼淺6裂，花瓣6枚，邊緣波浪狀；果實為蒴果，成熟後與花萼筒癒合成長橢圓柱狀。僅分布在台灣南部沿海珊瑚礁地區；蘭嶼則見於東部和北部。

林投 *Pandanus odoratissimus* L. f.
露兜樹科

多年生常綠性半蔓藤狀的大型灌木，氣生根很多；葉子螺旋狀排列方式聚生在枝條的頂端，除了葉緣有硬鋸齒外，葉背面中肋也是有硬鋸齒。雌雄異株且是屬於佛焰花序：雄花穗往往是下垂，雌花穗則是直立。果實由多枚有稜角的乾果聚集在一起，往往是圓球形；隨著果實的成熟，顏色也會由橄欖綠轉為橘紅色，這時往往因為太重而下垂。全台灣海岸珊瑚礁邊可以發現；蘭嶼全島也可以見到。

攝影／楊宗愈

白樹仔（*Gelonium aequoreum* Hance）、假三腳虌（*Melicope triphylla* (Lam.) Merr.）、草野氏冬青、琉球黃楊（*Buxus liukiuensis* Makino）、蘭嶼海桐（*Pittosporum moluccanum* Miq.）、基尖葉野牡丹（*Melastoma affine* D. Don）、密脈赤楠（*Syzygium densinervium* Merr. var. *insulare* C.E. Chang）、蘭嶼赤楠（*S. simile* (Merr.) Merr.）、雨傘仔（*Ardisia cornudentata* Mez）、蘭嶼馬蹄花、毛玉葉金花（*Mussaaenda pubescens* Ait. f.）、琉球九節木（*Psychotria manillensis* Bartl. ex DC.）、草海桐（*Scaevola taccada* (Gaertner) Roxb.）等；藤本植物主要種類是山林投（*Freycinetia formosana* Hemsl.），其他尚有菲律賓南五味子（*Kadsura philippinensis* Elm.）、菲律賓胡椒、雀梅藤（*Sageretia thea* (Osbeck) Johnst.）、歐蔓（*Tylophora ovata* (Lindl.) Hook. ex Steud.）、蘭嶼念珠藤（*Alyxia insularis* Kanehira & Sasaki）、小錦蘭（*Anodendron affine* (Hook. & Arn.) Druce）、栓壁龍（*Psychotria serpens* L.）、平柄菝葜（*Heterosmilax japonica* Kunth）、印度鞭藤（*Flagellaria indica* L.）等；在林下或是邊緣較有陽光處，草本植物才生長較多，例如田代氏黃芩（*Scutellaria tashiroi* Hayata）、粗穗蛇菰（*Balanophora fungosa* J.R. & G. Forst.）、刀傷草（*Ixeridium laevigatum* (Blume) J.H. Pak & Kawano）、白花草（*Leucas chinensis* (Retz.) R. Br.）、野菰（*Aeginetia indica* L.）、闊葉麥門冬、束草（*Carex brunnea* Thunb.）、乾溝飄拂草等。整體說來，本植物群落的種類也是十分豐富，約在250種上下。

濕生植物群落

濕生植物群落（Swamp vegetation）主要是指生長在蘭嶼紅頭村天池的池中及其附近的植物群落而言，由於天池附近非常潮濕，常常都被濃霧籠罩著，所以生長在此的植物都必須能耐潮濕，且有些還得長期浸泡在水中仍能存活，植被相以喬木、灌木為主，林下或是林緣仍有一些草本植物。

天池的邊緣多是灌木或小喬木生長，大喬木主要長在較裡面一些，較易辨識的有蘭嶼筆筒樹（*Cyatha fenicis* Copel.）、腰果楠（*Dehaasia incrassata* (Jack.) Kosterm.）、大花赤楠（*Syzygium tripinnatum* (Blanco) Merr.）、賽赤楠、山欖、厚葉石斑木（*Rhaphiolepis indica* (L.) Lindl. ex Ker var. *umbellate* (Thunb. ex Murray) Ohashi）、豬母乳、尖尾葉長葉榕（*Ficus heteropleura* Blume）、三脈紫麻（*Oreocnide trinervis* (Wedd.) Miq.）、蘭嶼新木薑子（*Neolitsea villosa* (Blume) Merr.）、蘭嶼鏽

蘭嶼白脈根 *Lotus australis* Andr.

豆科

一年生匍匐性草本；莖肥厚粗圓呈暗紅色。葉子是由5小葉所組成的羽狀複葉，小葉形狀為倒卵披針狀，肉質；花白色腋生，一般多為4（-5）枚的傘形花序；果實筆直不彎曲的圓柱形，成熟呈褐色。種子為黑色腎臟形。在台灣地區僅分布於蘭嶼的東北角。

棋盤腳樹 *Barrintonia asiatica* (L.) Kurz.

玉蕊科

常綠性多分枝的中型喬木；葉多密集生於枝條頂端，倒卵形或長橢圓倒卵形，近無柄，兩面均光滑無毛；總狀花序直立頂生，長有花4~20朵，花萼筒淺綠2~3裂，宿存，花瓣潔白，花絲及花柱頂端鮮紅色，花藥金黃色；果實呈4稜狀的角錐形，果皮很厚不開裂。在台灣僅分布於恆春半島及蘭嶼。

攝影／楊宗愈

葉灰木（*Symplocos cochinchinensis* (Lour.) S. Moore var. *philippinensis* (Brand) Noot.）、錫蘭玉心花（*Tarenna zeylanica* Gaertn.）、蘭嶼椌木（*Dysoxylum cumingianum* C. DC.）、蘭嶼肉豆蔻（*Myristica ceylanica* A. DC. var. *cagayanensis* (Merr.) J. Sinclair）、大野牡丹（鏽葉野牡丹）（*Astronia formosana* Kanehira）、革葉羊角扭（*Memecylon lanceolatum* Blanco）、山黃梔、琉球雞屎樹（*Lasianthus fordii* Hance）、蘭嶼九節木（*Psychotria cephalophora* Merr.）、山檳榔（*Pinanga tashiroi* Hayata）等；林下草本或林內的藤蔓性植物，或是那些直接浸泡在水裡面的種類，下面將列舉一些比較特別或是較常見的，例如錫蘭七子蕨（*Helminthostachys zeylanica* (L.) Hook.）、全緣卷柏（*Selaginella delicatula* (Desv.) Alston）、帶狀瓶爾小草（*Ophioderma pendula* (L.) Presl）、紅果金粟蘭（*Sarcandra glabra* (Thunb.) Nakai）、食用樓梯草（*Elatostema acuteserratum* B.L. Shih & Y.P. Yang）、長箭葉蓼（*Polygonum hastatosagittatum* Makino）、蘭嶼千金藤（*Stephania merrillii* Diels）、紅葉藤、直立半插花（*Hemigraphis cumingiana* (Nees) F.-Vill.）、番仔林投（*Dracaena angustifolia* Roxb.）、船仔草（*Curculigo capitulata* (Lour.) Kuntze）、畦畔莎草（*Cyperus haspan* L.）、假淡竹葉（*Centotheca lappacea* (L.) Desv.）、蘭嶼落檐（蘭嶼芋）（*Schismatoglottis kotoensis* (Hayata) T.C. Huang, A. Hsiao & H.Y. Yeh）、呂宋月桃（*Alpinia flabellate* Ridley）、山月桃仔（*A. intermedia* Gagn.）、白鶴蘭（*Calanthe triplicata* (Willem.) Ames）、紅花石斛（*Dendrobium miyakei* Schltr.）等。

蘭嶼的天池是一個火山口，所以池面範圍會隨著雨水的多寡而有增減，由於此處多霧，再加上遮蔽良好，所以種類頗多，估計至少在300種以上；不過本植物群落與下一個森林植物群落，將會有許多重複的種類。

森林植物群落

森林植物群落（Forest vegetation）為蘭嶼島主要的植物社會，包括地被植物、草本植物、灌木、喬木、藤本植物、寄生植物與附生植物，其種類及數目都是最豐富的，亦有類似熱帶雨林的植群結構（如板根、支柱根、幹生花等）。除了那些只出現在前面四種植物群落的種類外，幾乎都會重複在本植物群落中，所以說森林植物群落的質與量都是最豐富的。不過整體看來，植物仍是以熱帶分布的

大葉山欖 *Palaquium formosanum* Hayata
山欖科

常綠性大喬木；葉片多簇生於枝條頂端，橢圓形或倒卵形，革質，葉柄很短；花朵腋生，簇生排列，淺黃綠色，花萼兩輪，每輪有裂片3枚；花瓣6枚，淺黃色；雄蕊 18~24 枚，花柱單一。果實呈橄欖球形漿果；種子一枚，暗褐色。台灣原生於北部及南部海濱；蘭嶼幾乎全島可見。

番龍眼 *Pometia pinnata* Forst.
無患子科

常綠性中型喬木，樹幹基部會形成「板根」；葉子是互生的一回奇數羽狀複葉，新生葉往往呈紅棕色；小葉長橢圓形或長披針形，厚紙質；圓錐花序腋生或頂生；花雜性花；花萼杯狀，花瓣圓形，雄蕊伸出花朵之外，花絲白色，花藥紅色，花柱延長。果實球形，不開裂且多肉的核果。種子棕褐色，被透明粘質的假種皮包裹著。在台灣僅分布於東部及蘭嶼。

攝影／楊宗愈

科、屬種類為主，例如桑科、蕁麻科、大戟科、楝科、桃金孃科、野牡丹科、山欖科、茜草科等等，當然那些全球廣泛分布的科（例如：豆科、菊科、禾本科、蘭科等），在本島的種類也很多，概括地估計，本植物群落應該至少有600種以上。

由於本植物群落的種類實在是太多，以下試著依照科別舉出一些常見的種類。桑科植物：麵包樹、垂榕（白榕）（*Ficus benjamina* L.）、大葉雀榕、對葉榕、牛奶榕（*F. erecta* Thunb. var. *beecheyana* (Hook. & Arn.) King）、豬母乳、澀葉榕、厚葉榕、蔓榕、鵝鑾鼻蔓榕、綠島榕（*F. pubinervis* Blume）、蘭嶼落葉榕（*F. ruficaulis* Merr. var. *antaoensis* (Hayata) Hatusima & J.C. Liao）、大冇榕、山豬枷、越橘葉蔓榕（*F. vaccinioides* Hemsl. ex King）、幹花榕（*F. variegata* Blume var. *garciae* (Elm.) Corner）、白肉榕及小葉桑等；蕁麻科植物有：密花苧麻（*Boehmeria densiflora* Hook. & Arn.）、青苧麻（*B. nivea* (L.) Gaudich. var. *tenacissima* (Gaudich.) Miq.）、瘤冠麻（*Cypholophus moluccanus* (Blume) Miq.）、紅頭咬人狗（*Dendrocnide kotoensis* (Hayata ex Yamam.) B.L. Shih & Y.P. Yang）、食用樓梯草、糯米團（*Gonostegia hirta* (Blume) Miq.）、四脈麻、蘭嶼水絲麻（*Maoutia setosa* Wedd.）、三脈紫麻、落尾麻（*Pipturus arborescens* (Link) C. Robinson）等；大戟科植物有：紅頭鐵莧、花蓮鐵莧（*Acalypha suirenbiensis* Yamam.）、茄苳（*Bischofia javanica* Blume）、鐵色、蘭嶼土沈香、密花白飯樹（*Flueggea virosa* (Roxb. ex Willd.) Voigt）、紅肉橙蘭（*Macaranga sinensis* (Baill.) Muell.-Arg.）、血桐（*M. tanarius* (L.) Muell.-Arg.）、野桐（*Mallotus japonicus* (Thunb.) Muell.-Arg.）、粗糠柴（*M. philippinensis* (Lam.) Muell.-Arg.）、蟲屎（*Melanolepis multiglandulosa* (Reinw.) Reich. f. & Zoll.）、圓葉血桐（*Omalanthus fastuosus* F.-Vill.）等；芸香科植物有：台灣香檬（*Citrus depress* Hayata）、假三腳鼈、長果月橘（*Murraya paniculata* (L.) Jack. var. *omphalocarpa* (Hayata) Swingle）、蘭嶼花椒（*Zanthoxylum intergrifoliolum* (Merr.) Merr.）等；楝科植物有：蘭嶼樹蘭、大葉樹蘭、穗花樹蘭（*Aphanamixis polystachya* (Wall.) R.N. Parker）、蘭嶼擬堅木（*Chisocheton patens* Blume）、蘭嶼堅木（*Dysoxylum arborescens* (Blume) Miq.）、蘭嶼椌木、大花椌木（*D. parasiticum* (Osbeck) Kosterm.）等；桃金孃科植物有：賽赤楠、密脈赤楠、蘭嶼赤楠、台灣棒花蒲桃（*Syzygium taiwanicum* C.E. Chang & Miau）、大花赤楠等；野牡丹科的植物有：大野牡丹、蘭嶼野牡丹藤（*Medinilla hayataina* H. Keng）、基尖葉野牡丹、革葉羊角扭等；山欖科的植物有：

蘭嶼羅漢松 *Podocarpus costalis* Presl
羅漢松科

常綠性低矮灌叢或小喬木；葉叢生於枝條頂端，線狀倒卵披針形，革質，雙面均光滑無毛；雌雄異株；雄花穗單生無柄，淺黃色，雌花單一，淺灰綠色；種子橢圓形至球形，長在一深紅色肉質圓柱狀的果托上，成熟時為紫黑色。在台灣地區僅分布於蘭嶼。

山菊 *Farfugium japonicum* (L.) Kitam.
菊科

多年生具有短根莖的草本植物；基生葉肉質，腎臟形，淺凹齒狀葉緣，幾乎兩面全光滑無毛；具有長花梗的繖房花序；頭狀花大型，直徑約4~6公分長，金黃色；瘦果細長，被白色長毛。在台灣僅分布於蘭嶼和綠島地區。

攝影／楊宗愈

大葉山欖、蘭嶼山欖及山欖等；茜草科植物：水冠草(*Argostemma stolaniflorum* Elmer)、山黃梔、苞花蔓(*Geophila herbacea* (Jacq.) O. Ktze.)、葛塔德木、小仙丹花(*Ixora philippinensis* Merr.)、雞屎樹(*Lasianthus obliquinervis* Merr.)、圓葉雞屎樹(*L. wallichii* Wight)、檄樹、大葉玉葉金花(*Mussaenda macrophylla* Wall.)、毛玉葉金花、欖仁舅(*Neonauclea reticulata* (Havil.) Merr.)、小花蛇根草(*Ophiorrhiza kuroiwae* Makino)、蘭嶼九節木、琉球九節木、拎壁龍、錫蘭玉心花、貝木(*Timonius arboreus* Elmer)、恆春鉤藤(*Uncaria lanosa* Wall. var. *appendiculata* (Benth.) C.E. Ridsdale)、呂宋水錦樹等；而豆科、菊科及禾本科等泛世界分布的植物，主要分布在全島的森林邊緣及一些向陽的地方，當然森林內也有一些草本或是蔓藤性的種類；蘭科植物則多是在森林內，有土生、樹生和腐生的種類，然多是以熱帶區系的種類為主。

耕作植物群落

在蘭嶼流行一句話：「會動的不關，不會動的要關起來」。原來島上的原住民對所飼養的豬和羊是採放牧型，為了避免牠們去啃食所種植的作物，只有將植物「關」起來嘍！

島上居民的主食是芋頭，所以放火燒山後開墾種植的主要作物就是水芋(*Colocasia esculenta* (L.) Schott)；而在一些旱地或屋舍旁，則會種植檳榔(*Areca catechu* L.)、椰子(*Cocos nucifera* L.)、地瓜(蕃薯)(*Ipomoea batatas* (L.) Lam.)、千年芋(里芋)(*Xanthosoma sagittifolium* (L.) Schott)、小米(*Setaria italica* (L.) Beauv.)、大薯(*Dioscorea alata* L.)、刺薯蕷(*D. esculenta* (Lour.) Burk. var. *spinuosa* (Roxb.) Kunth)及木瓜(*Carica papaya* L.)等。

遷移的重要橋樑

蘭嶼島之維管束植物歷經前人多年來的研究統計大約有800~850種，其中約有650種分布於台灣，約有500種也出現在菲律賓，可見本島的植物受台灣影響較大；然而若僅就木本植而言，就有110種僅見於菲律賓而不見於台灣，所以認為蘭嶼地區的木本植物區系又受菲律賓影響較台灣為大；然以植物分布的科屬言之，蘭嶼島的熱帶區系成分確實比較重。綜合上述，可以看出本島上的植物確實具有很強烈的「交匯地帶」的特色；若再以植物地理來看，蘭嶼的確可以說是台灣與菲律賓的交匯點，也很可能是過去植物遷移的一個重要橋樑。

青脆枝 *Nothapodytes nimmoniana* (Graham) Mablerley
茶茱萸科

常綠性小喬木；葉片互生呈螺旋狀排列，橢圓形、長橢圓形或長橢圓披針形，光滑無毛的紙質，具長葉柄；頂生的聚繖花序或是繖房花序；花萼杯狀5裂，花瓣5枚，淺黃色，雄蕊5枚，與花瓣等長且互生；果實呈卵形或橢圓形的核果，成熟時為暗紫紅色；種子一枚。在台灣僅出現於蘭嶼及綠島的林緣地區。

蘭嶼蘋婆 *Sterculia cerramica* R. Br.
梧桐科

多年生的中型常綠性喬木；單葉集中在枝條頂端且以螺旋狀方式排列生長，窄心形；圓錐花序頂生，且為雌雄同株的雜性花；由於蘋婆屬的植物都沒有花瓣，乃由黃綠色萼片組成鐘形花；果實成熟時為橙紅色且會開裂露出裡面黑色的種子。在台灣僅分布蘭嶼及綠島的林緣地區。

攝影／楊宗愈

粗莖麝香百合 *Lilium longiflorum* Thunb. var. *scabrum* Masam.

百合科

可以長到60~90公分高的多年生草本植物，莖上有粗糙的毛茸；花純白色頂生，往往同時開放三、四朵；花凋謝後剩下棒狀的子房，漸漸發育生長成為圓柱狀的蒴果，最後轉為成熟的金黃色，稍加震動，就會開裂散播出暗褐色種子。台灣特有變種，分布於全島海邊及蘭嶼地區。

四脈麻 *Leucosyke quadrinervia* C. Robinson

蕁麻科

約2~7公尺高的常綠小喬木或灌木；葉子互生，歪斜狀的卵形或橢圓形，雖然葉脈是3~5條主葉脈，但葉子基部有些歪斜，所以看起來像僅有四條主葉脈；托葉長披針形或長三角形；雌雄異株，雌花和雄花小，腋生；果實瘦果，圓球形的頭狀花序。在台灣地區僅出現在蘭嶼及綠島。

密脈赤楠 *Syzygium densinervium* Merr. var. *insulare* C.E. Chang

桃金孃科

常綠性小喬木或灌木；葉對生倒卵狀長橢圓形，革質，側脈多數且相互平行；頂生的繖形花序，花萼筒紡錘形，光滑無毛，花瓣白、小，與萼筒相連，花絲多數白色，開展生長；果實卵狀橢圓形，暗紅色。台灣特有變種，僅分布於屏東縣及蘭嶼地區。

錫蘭七指蕨

Helminthostachys zeylanica (L.) Hook.

瓶爾小草科

根莖肉質粗短的直立蕨類；苞子囊穗單一直立，頂生於成熟的葉柄上；葉子基本上三出，而每一部份又再呈2~3披針形裂片；子囊穗具有長柄，每一苞子囊簇生在一極短的小枝條上。在台灣，僅在蘭嶼天池的附近被發現過。

腰果楠 *Dehaasia incrassata* (Jack.) Kosterm.

樟科

常綠中型喬木；葉墨綠色互生，光滑無毛，橢圓形或卵狀長橢圓形；腋生圓錐花序，花小淺黃綠色；小花梗宿存且發育呈紅色肉質圓柱狀，下端才是橢圓形的黑色核果。在台灣僅分布蘭嶼森林之中。

欖仁舅 *Neonauclea reticulata* (Havil.) Merr.

茜草科

常綠大型喬木；單葉對生幾乎無柄，倒卵形或闊倒卵形，兩面均光滑無毛；花呈密生球形的頭狀花序，單一或三枚叢生，白色；果實密集排列成球形頭狀，直徑可達 3.5 公分；種子多枚，扁平有翅。在台灣僅分布於東部山區或海濱；蘭嶼見於山區及森林邊緣。

蘭嶼野牡丹藤 *Medinilla hayataina* H. Keng

野牡丹科

藤蔓性的灌木；葉子單葉輪生，長橢圓形，三出葉脈且葉緣全緣；花序為腋生的繖形圓錐花序，花呈淺粉紅色，雄蕊有 8枚，等長，花絲白色花藥藍紫色；果實漿果，成熟時為暗紅到紫黑色。為蘭嶼特有種，目前僅知道分布於紅頭山區。

水芋 *Colocasia esculenta* (L.) Schott

天南星科

原產於熱帶亞洲地區，但卻在熱帶及亞熱帶地區廣泛被種植；葉闊卵形或略呈圓形，葉基部心形，葉柄淺綠色或紫色。當地人稱這類種在水中的芋頭為紅山芋或香芋，蘭嶼話為「uradede」，也是當地人主要食用者；稱陸地上生長的千年芋（山芋）為「furadede」；而姑婆芋「raum」及蘭嶼落檐「authe」的葉子，在蘭嶼只有當盤子用嘍。

攝影／楊宗愈

攝影／桂曉芬

作者簡介(按文章順序排列)

郭城孟 Chen-Meng Kuo

學歷：瑞士蘇黎世大學系統植物學研究所／博士 1981/09～1984/08
現職：國立台灣大學植物系、所／副教授 1985/08～迄今
國立台灣大學植物系標本館／館長 2000/08～迄今

細緻而多樣化是台灣的生態特色，台灣的每一個角落，都是與眾不同的。面對這樣不可多得的資產，我們應該是想怎麼可以世世代代保有她，而不是一心想著可以變賣多少錢。

彭鏡毅 Ching-I Peng

學歷：國立中興大學植物系／學士 1968/09～1972/06
國立台灣大學植物研究所／碩士 1974/09～1976/06
美國華盛頓大學生物系／博士 1978/08～1982/05
經歷：中央研究院植物研究所／研究助理 1978/01～1978/07
中央研究院植物研究所／副研究員 1982/12～1987/07
國立自然科學博物館／學術副館長 1995/11～1997/12
國立自然科學博物館／代理館長 1996/09～1997/10
現職：中央研究院植物研究所／研究員 1987/08～迄今

台灣，這個僅佔全球陸地面積0.02%的蕞爾小島，孕育著全球1.5~2%生物多樣性，其中大約四分之一為特有物種。謂之寶島，誰曰不宜？

楊遠波 Yuen-Po Yang

學歷：美國聖路易大學生物系／博士 1983～1988
經歷：台灣省林業試驗所森林生物系／副研究員 1989～1992
國立中山大學生物科學系／副教授 1992～1999
現職：國立中山大學生物科學系／教授 1999～迄今

生長在台灣的每一個人都有責任保育台灣的自然環境與資源，而不是僅賴少數人的努力，希望台灣的自然資源保育工作一天比一天的好。

薛美莉 Mai-Li Hsueh

學歷：國立中興大學森林學研究所／碩士 1987/09～1990/06
經歷：林業試驗所／助理 1991/08～1992/07
現職：特有生物研究保育中心解説教育組／副研究員 1992/07～迄今

剛開始投入台灣生態研究時，充滿了憧憬，然而在現實面看到美麗的小島正一吋吋的被分割，感到萬分無奈。幸好近來一些鄉土省思，讓許多非專業人士投入到這個領域，讓人覺得台灣的生態依然是充滿希望。

曾喜育 Hsy-Yu Tzeng

學歷：國立中興大學森林學系／學士 1993/09～1997/06
國立中興大學森林學研究所／碩士 1997/09～1999/06
經歷：林業試驗所福山分所／助理 1999/12～2000/11
現職：林業試驗所恆春分所／助理 2000/11～迄今

伍淑惠 Shu-Hui Wu

學歷：私立中國文化大學森林學系／學士 1993/09～1996/06
國立台灣大學森林學研究所／碩士 1996/09～1999/01
經歷：中央研究院植物研究所／研究助理 1999/02～1999/11
現職：林業試驗所恆春分所／助理 1999/12～迄今

邱文良 Wen-Liang Chiou

學歷：美國愛荷華州立大學植物系／博士 1992～1996
經歷：林業試驗所恆春分所／助理研究員 1980～1990
林業試驗所生物系／助理研究員 1990～1998
林業試驗所生物系／副研究員 1998～2000
現職：林業試驗所福山分所／副研究員 2000～迄今

學習與自然相處，將使個人心胸開闊，社會環境詳和。

楊國禎 Kuoh-Cheng Yang

學歷：國立台灣大學植物學系／學士 1979/09～1983/07
國立台灣大學植物學研究所／碩士 1985/09～1988/07
國立台灣大學植物學研究所／博士 1990/09～1996/07
經歷：國立台灣大學植物學系／助教 1985/08～1992/07
台灣省林業試驗所恆春分所／助理 1994/06～1995/02
台灣省林業試驗所生物系／助理 1995/02～1997/07
私立靜宜大學人文科／兼任副教授 1996/09～1997/06
私立靜宜大學人文科／副教授 1997/07～2001/07
現職：私立靜宜大學生態學研究所／副教授 2001/08～迄今

國人太重視自身的短暫利益，幾乎完全忽視人與環境的關係，骨子裡還信奉人是萬能的、人定可以勝天的教條，這樣的想法放在現在時、空背景下的台灣，如不改變，災難將永續不絕！

謝長富 Chang-Fu Hsieh

學歷：國立台灣大學植物學研究所／博士 1981/08
現職：國立台灣大學植物系、植物研究所／教授 1982/08～迄今

山林是否有情，端視人們如何看待自然，想要回歸自然，首先必先學會尊重自然，疼惜自然，大地才會接受你。

陳子英 Tze-Ying Chen

學歷：私立中國文化大學森林系／學士 1978/09～1982/06
國立台灣大學森林系／碩士 1986/09～1988/06
國立台灣大學森林系／博士 1989/09～1994/06
經歷：茶業改良場台東分場茶作課／助理研究員 1984/07～1985/06
宜蘭農工專科學校森林科／講師 1992/08～1994/08
現職：宜蘭技術學院森林系／副教授 1994/08～迄今

台灣的生態環境是美好的，
但未來應在保育的前提下，
在合理的發展與妥善的保存之間，
取得動態的平衡。

黃曜謀 Yao-Moan Huang

學歷：私立文化大學森林系／學士 1988/10～1992/06
國立台灣大學森林所／碩士 1992/09～1995/06
現職：林業試驗所生物系／約僱助理 2000/08～迄今

莫讓今日的美景成為明日的回憶！

呂勝由 Sheng-You Lu

學歷：國立台灣大學森林所／碩士 1990～1992
現職：林業試驗所生物系／助理研究員 1991～迄今

大家都說新加坡很美，據瞭解，新加坡已經沒有天然林了，因此其展現的只是現代人文環境中的人工美。台灣百分之五十八為森林覆蓋，在自然環境中，我們擁有天然林盡情表露的自然之美，是無可取代的。天然林一但砍伐就難以再回復，因此值得我們大家的珍惜、愛護。

張藝翰 Yih-Hann Chang

學歷：國立中興大學植物學系／學士 1990/09～1994/07
　　　國立台灣大學植物學研究所／碩士 1994/09～1998/07
現職：林務局東勢林管處梨山工作站／技術助理員 2001/11～迄今

人類利用其天賦，得以適存於天地之間、凌駕萬物之上，卻也因此淡忘其源於自然。
學習對大地謙卑與對萬物生命之尊重，除了將良善我們的內心，也能豐富我們的生命。

吳瑞娥 Juei-Er Wu

學歷：私立中國文化大學森林系／學士 1990/09～1993/06
　　　私立中國文化大學生物科技所／碩士 1993/09～1995/06
　　　國立台灣師範大學生物所／博士 1995/09～2001/06
經歷：台北縣立忠孝國中自然科／教師 2000/08～2001/07
現職：台北市立明德國中自然科／教師 2001/08～迄今

一個有錢的社會，若失去對自然、生命的關懷和尊重，冷漠與失序將是無可避免。然而若人人能兼持「人養地，地養人」的生活哲學，學習關懷生命、尊重自然，將能使生活更快樂。

鄭育斌 Yu-Pin Cheng

學歷：國立台灣大學植物系／學士 1986/09～1990/06
　　　國立台灣大學植物所／碩士 1990/09～1992/06
　　　國立台灣大學植物所／博士班 2000/09～迄今
經歷：林業試驗所恆春分所／助理 1996/04～1998/01
　　　林業試驗所生物系／助理 1998/02～2001/10
現職：林業試驗所生物系／助理研究員 2001/10～迄今

讓台灣美好的事物和美麗的環境能生生不息，永遠存在......

曾彥學 Yen-Hsueh Tseng

學歷：國立台灣大學森林研究所／碩士 1989/07～1991/06
　　　國立台灣大學森林研究所／博士班 1999/09～迄今
經歷：特有生物研究保育中心植物組／助理研究員 1992/07～2001/10
現職：特有生物研究保育中心植物組／副研究員 2001/11～迄今

為萬萬物物留下一線永續生機；
為子子孫孫留下一片美好樂土。
——保育遠景

徐嘉君 Chia-Chun Hsu

學歷：國立台灣大學植物研究所／碩士 1996/09～1998/06
經歷：林務局台東林管處知本工作站／技術員 2000/12～2001/04
　　　林業試驗所六龜分所／研究助理 2001/04～2002/03
現職：林業試驗所／研究助理 2002/04～迄今

親近自然，使人與環境之間不再有疏離感。

邱少婷 Shau-Ting Chiu

學歷：國立台灣大學植物學系／學士 1979/09～1983/06
　　　國立台灣大學植物學系／碩士 1983/09～1986/06
　　　美國密西根州立大學植物與植物病理學系／博士 1986/09～1992/02
經歷：私立東海大學生物學系／兼任副教授
現職：私立靜宜大學生態學研究所／兼任副教授
　　　國立台灣大學植物學所／兼任副教授
　　　國立自然科學博物館植物學組／副研究員

企盼生命往前邁進的過程中，
學習尊重與被尊重的自然平衡，
共享生存處所的點滴資源，
再現福爾摩沙的璀璨。

劉靜榆 Ching-Yu Liou

學歷：國立台灣大學森林學研究所／碩士 1989/09～1991/06
　　　國立台灣大學森林學研究所／博士班 1999/09～迄今
經歷：玉山國家公園管理處保育課、解說課／技士 1991/08～1992/07
現職：特有生物研究保育中心棲地生態組／助理研究員 1992/07～迄今

每一種生物都是經過長久的演化，在自然界扮演不同的角色，我們可以感覺到有許多物種的族群正在迅速的消減，其實人們許多的享用都來自大自然中許多生命的犧牲。

賴國祥 Kwo-Shang Lai

學歷：國立中興大學森林學系／學士 1977/09～1981/06
　　　國立中興大學森林研究所／碩士 1981/09～1983/06
　　　國立中興大學植物學研究所／博士 1988/09～1992/01
經歷：國立中央大學理學院／助教 1985/10～1992/07
　　　特有生物研究保育中心棲地生態組／副研究員 1992/07～1993/09
現職：特有生物研究保育中心高海拔試驗站／副研究員兼站主任 1993/09～迄今

學了這些年的生態學，總覺得自然有它自己的道理，人只要靜靜的看和學習如何與自然相處就可以了，因為常常最好的做法，就是什麼都不要做。

王震哲 Jenn-Che Wang

學歷：國立台灣師範大學生物學系／學士 1971/09～1976/06
　　　國立台灣大學植物學研究所／碩士 1976/09～1978/06
　　　國立台灣大學植物學研究所／博士 1984/09～1988/06
經歷：國立台灣師範大學生物系／副教授 1989/07～1996/07
現職：國立台灣師範大學生物系／教授 1996/08～迄今

用心去體會台灣之美，用行動去實踐對台灣的愛。

韓中梅 Chung-May Han

學歷：國立台灣師範大學生物學系／學士 1994/08～1998/07
　　　國立台灣師範大學生物學系／碩士 1999/08～2001/07
現職：國立台灣師範大學生物學系／助教 2001/08～迄今

沒有豔麗外表的生物往往才是舉足輕重的生態基石，明星物種的保護傘是立在這些傻傻的物種上，如何著手開始瞭解這些不起眼的生命，是目前的重要課題。

陳志雄 Chih-Hsiung Chen

學歷：私立輔仁大學生物系／學士 1986/09～1991/07
　　　國立台灣師範大學生物學研究所／碩士 1994/09～1996/07
　　　國立台灣師範大學生物學研究所／博士 1996/09～2001/07

大自然的美，是不吝嗇與他人分享的最大財富；而生態環境的破壞，對自己、對後代，都是最無奈的悲情。希望大家盡力，不要空留回憶。

楊宗愈 T.Y. Aleck Yang

學歷：私立文化大學植物系 1980/10～1981/06
　　　私立東海大學生物系／學士 1981/09～1984/06
　　　國立台灣大學植物所／碩士 1984/09～1987/06
　　　英國瑞丁大學植物系／博士 1989/10～1994/12
經歷：私立文化大學景觀學系／副教授 1995/09～2001/01
現職：國立自然科學博物館植物學組／副研究員 1995/09～迄今
　　　私立東海大學生物學系／副教授 2001/02～迄今

要能關心自然，必須要瞭解自然；要想認識台灣生態及環境，請先熟悉我們周遭的一草一木，還請先從我們做起。

國家圖書館出版品預行編目資料

發現綠色台灣：台灣植物專輯／郭城孟主編．
初版．--臺北市：農委會林務局，民91

面；　公分

ISBN 957-01-1232-8（精裝）

1. 植物 — 臺灣 — 圖錄

375.232　　91011343

發現綠色台灣

台灣植物專輯

出版者	行政院農業委員會林務局 社團法人中華民國企業永續發展協會
發行人	黃裕星、陳耀生
總策劃	吳淑華、黃正忠
執行	蕭裕陸
編輯顧問	郭城孟、楊遠波、劉和義、彭鏡毅、郭長生
編輯策劃	台灣大學植物系植物標本館、林文集
主編	郭城孟
執行編輯	高美芳、賴郁旻
美術設計	桂曉芬、李健邦
贊助單位	行政院國家永續發展委員會 行政院環境保護署 中美和文教基金會 中國石油公司 台積電文教基金會
發行所	行政院農業委員會林務局 台北市杭州南路一段2號 電話：(02)2351-5441 社團法人中華民國企業永續發展協會 台北市承德路一段70-1號8樓之1 電話：(02)2550-1792 傳真：(02)2550-6309
出版日期	民國九十一年六月初版一刷
定價	新台幣 2000元
設計製作	高遠文化事業有限公司　(02)2751-7911
製版印刷	五洲製版印刷股份有限公司　(02)2880-6598

GPN：1009101712
ISBN：957-01-1232-8